T0299293

Re-envisioning Organizations through Transformational Change

The journey towards the future of work was greatly accelerated due to the COVID pandemic. Some changes have altered the functioning of the business world forever. Against the backdrop of these alterations, variations, and modifications, this book presents and analyzes three crucial factors: work, workforce, and workplace and their transformation into new-age organizations for meeting its customer expectations and long-term strategic goals. Companies must focus on ways of deployment of policies and practices that meet the business needs from the perspective of external changes. To achieve this goal, the organizations must realign their stakeholders and indulge in critical thinking by looking deeply into factors responsible for bringing about this transformational change. Re-envisioning is the current critical need for organizations to thrive; they must incorporate best practices to beat the competition and add value to their existing HR processes. This book clearly presents the practices and policies of successful organizations through the contribution of industry leaders.

This book helps you understand the dynamism of work, workforce, and workplace that exist in organizations (as well as the challenges these organizations face) and their impact on business practices. The authors cover these broad areas because of the need to diversify and promote organic inclusive growth.

Essentially, re-envisioning our organizations is the new normal. Organizations must leave the shackles of what might have been and look to what they can be. Stakeholders, employees, and the environment have been drastically altered, and organizations must change accordingly to survive. What now matters is how much an organization re-envisions itself and how it deals with all that is happening.

Re-envisioning Organizations through Transformational Change

A Practitioners Guide to Work, Workforce, and Workplace

Edited By

Poornima Madan

Shruti Tripathi

Fehmina Khalique

Geetika Puri

A PRODUCTIVITY PRESS BOOK

First published 2023
by Routledge
605 Third Avenue, New York, NY 10158

and by Routledge
4 Park Square, Milton Park, Abingdon, Oxon, OX14 4RN

Routledge is an imprint of the Taylor & Francis Group, an informa business

ISBN: 978-1-032-21308-8 (hbk)
ISBN: 978-1-032-21306-4 (pbk)
ISBN: 978-1-003-26775-1 (ebk)

DOI: 10.4324/9781003267751

Typeset in Garamond
by KnowledgeWorks Global Ltd.

Contents

Preface

Organizations are often seen as entities that requires the key stakeholders to maximize profit margins, increase brand image, improve the quality of products/services for customer satisfaction, and enhance employee morale. Therefore, it becomes imperative to redesign work keeping in mind the key stakeholders, like customers, investors, and employees. Since the world is still emerging from the global crisis of the pandemic and the challenges posed by the new world order, both organizations, as well as individuals, are on a constant lookout for ways to sustain business. Having said this, change seems to be the only constant for the organizations and their stakeholders who stand tall today. Hence, the need for transformational change arises in order to beat the competition and add value to the business for strategic goals achievement.

Globalization and digital transformation have taught businesses the importance of innovation in all spheres whether it is products or services, workplace culture, customer relations, or talent retention. Needless to say, the global businesses across the geographies are technology-driven, and revamping the organizations has become a necessary evil today. Therefore, there lies an ardent need to re-envision the ways businesses formulate and execute their plans and policies.

There has been a need felt among researchers and academicians for a research-focus compendium of knowledge that would help the practicing managers and new researchers in understanding the ways organizations transform their policies and practices in order to flourish in the dynamic competitive world. Several researchers in the past have focused on the ever-changing dynamic environment of business; however, the sea-change brought due to pandemics in the business world is still left unexplored up to a large extent.

The current book encompasses work, workforce, and workplace, and their transformation into new-age organizations in order to meet the challenges posed by the deadly COVID-19 pandemic. The primary aim is to focus on ways of deployment of policies and practices that meet the business needs from the perspective of economic changes. To achieve this goal, the organizations need to realign their stakeholders and indulge in critical thinking by looking deeply into the factors responsible for bringing about transformational change.

This book will make an effort to suggest dynamism and challenges in the work, workforce, and workplace that the organizations across the globe are bringing and their impact on the business practices. Our justification to cover these broad areas is based on the need to diversify and promote organic inclusive growth. This book is an effort by the editors to highlight the changes that have been made and the difference it has brought to the organization. The chapters will highlight what is new in the organization in terms of its work practices, workforce practices, and overall workplace, especially post-pandemic. Moreover, it will address the problem of finding interdisciplinary cumulative research for its audience along with offering a unique perspective, where a microcosm of knowledge will be available to the readers in one book.

This book would not have been possible without the collaborative efforts and support of practicing managers in the corporate world, academicians, and researchers who joined hands with us in this project. Our contributors have relentlessly worked to provide input on the current business scenario through their chapters. They revised their drafts several times to improve the quality of content in their chapters in line with feedback provided by the reviewers and editors. We are thankful to our contributors for meeting the timelines and devoting their efforts and time in order to make this book a success. We are thankful to Taylor & Francis for their commitment and support in publishing this volume. A special mention to Mr. Michael Sinochi (Publisher-Productivity Press) for hand holding the editors right from the inception to the submission of the manuscript. Without the support of these major stakeholders, this dream could not have been realized.

We are extremely grateful to Dr. Rohan Rozario, Head-Organizational Development and Change, Public Investment Fund (Saudi Arabia's sovereign wealth fund) in the United Kingdom, for the "Foreword" he has so graciously written for the volume.

Last but not the least, we would like to express our gratitude to our family members who continuously supported and encouraged us on this journey.

Dr. Poornima Madan
Assistant Professor OB & HR,
Jaipuria Institute of Management, Noida

Dr. Shruti Tripathi
Professor, School of Employability and Holistic Development,
Delhi Skill and Entrepreneurship University, Delhi, India

Dr. Fehmina Khalique
Associate Professor OB & HR, Lloyd Business School,
Greater Noida, India

Ms. Geetika Puri
Director, Philomath Research Pvt; Ltd.,
New Delhi, India

Foreword

A majority of organizations across the globe are facing an existential crisis resulting from the COVID-19 pandemic, leading to an urgent need to transform their workforce, workplace, and management processes. The corporate world has witnessed the workforce being dislocated, supply chains seeing a massive shift, and a high rate of unemployment, which has resulted in changing perspectives of work (Amis and Greenwood 2020). Owing to this, organizations need to rethink and design work practices that are sustainable and enable them to be competitive in the new work environment.

Since the pandemic brought unprecedented challenges, many organizations are now designing in advance as part of crisis management (Mikušová and Horváthová 2019), also creating working conditions that are more conducive to protecting the employees from unforeseen events (Gartner, 2020; KPMG, 2020) As per a survey by McKinsey (2020), organizations are trying to redesign their workplace to generate not only secure but also rewarding careers for employees. The pandemic has taught a lesson to organizations about how the virtual platforms have played a key role in the continuity of business thereby challenging the brick-and-mortar model (Gartner, 2020)

Keeping the workforce engaged and productive is a key success factor for any organization. This is going to be the biggest challenge in the workplace today. Organizations are now constantly exploring new ways of working to do this. The new normal has made organizations segregate the extent of their activities into face-to-face ones and virtual ones depending on their needs and expectations (BCG 2020; PricewaterhouseCoopers 2020). As per World Economic Forum (2020), it is inevitable for the organizations to ignore the segregation of activities in the workplace as face-to-face ones, virtual ones, or amalgamation of both as a successful practice in order to thrive for cutting edge. In a similar survey published by Harvard Business Review (2020), it was pointed out that it could be advantageous for the organizations as well as employees to design their workplace where the business operated partially virtual and partially physical; this gives leverage to employees to work remotely or join the physical office lessening the geographical constraints faced earlier. This also calls for redesigning the jobs as per the availability of employees. Consequently, employees can live at a place of their choice, either with their families or in areas with a low cost of living. This will lead to higher employee engagement and workplace performance improving the overall productivity of companies (BCG 2020).

In addition, given the costs of office space, optimization of the workspace is important. The question of how much space shall be used post-work-from-home mode of working to physical workstations poses many challenges. The answer lies in optimum utilization of resources, for example, choosing the location for the office and other modalities like rent, ownership, hybrid work, etc. When designing and revamping the workplace, the employees' physical fitness should be an important consideration; if neglected it can result in serious illness, absenteeism from work, demotivation, poor performance, etc. (Pfeffer 2010). The need of the hour is to rethink the inhuman design of the physical workplace into a more conducive environment that can engage employees in meaningful assignments (Michaelson et al., 2014).

Organizations must find innovative ways to develop job duties and responsibilities in a manner that matches the new normal depending on the physical and virtual requirements of the tasks. These organizations need to realign their workforce into innovative processes and technologies for achieving long-term strategic goals. For example, the office workplace must comply with employees' safety, well-being, technological advancements, hybrid mode of work, and low hierarchy with more flexibility (Ancillo et al., 2020).

A key consideration to be aware of for any workplace transformation and virtual working to be successful is it requires a certain discipline, capability, mindset, and culture change within the workforce. One size does not fit all, and the biggest mistake an organization could make is to try to follow others, without really understanding what is best for them. Management of employees and activities in a virtual world is a lot more complex than the traditional face-to-face office environment. The transition is not easy and takes time. It is also critical that this transition to new ways of working is supported by the right technology and management processes, aligned to the strategy of the organization.

Dr. Rohan de Rozario
Head – Organizational Development and Change,
Public Investment Fund, Saudi Arabia

References

Amis, John, and Royston Greenwood. 2020. "Organisational Change in a (post-) Pandemic World: Rediscovering Interests and Values." *Journal of Management Studies*. doi:10.1111/joms.12663.

BCG. 2020. Remote work works – Where do we go from here? BCG. https://www.bcg.com/publications/2020/remote-work-works-so-where-do-we-go-from-here.

de Lucas Ancillo, Antonio, María Teresa del Val Núñez, and Sorin Gavrila Gavrila. 2021. "Workplace Change within the COVID-19 Context: A Grounded Theory Approach." *Economic Research-Ekonomska Istraživanja* Vol. 34 (1): 2297–2316. doi:10.1080/1331677X.2020.1862689.

Gartner. 2020. Future of work trends post-COVID-19. Gartner. https://www.gartner.com/smarterwithgartner/9-future-of-work-trends-post-covid-19/.

Harvard Business Review. 2020. Workforce strategies for post-COVID-19 recovery. Harvard Business Review. https://hbr.org/sponsored/2020/05/workforce-strategies-for-post-covid-19-recovery.

KPMG. 2020. *COVID-19 Returning to the new reality – Return to workplace*. https://home.kpmg/au/en/home/insights/2020/05/coronavirus-covid-19-return-to-workplace.html.

McKinsey. 2020. Reimagining the Office and Work Life after COVID-19. Mckinsey & Company, June, 1–5. https://www.mckinsey.com/business-functions/organization/our-insights/reimagining-the-office-and-work-life-after-covid-19.

Michaelson, C., M. G. Pratt, A. M. Grant, and C. P. Dunn. 2014. "Meaningful Work: Connecting Business Ethics and Organization Studies." *Journal of Business Ethics* Vol. 121(1): 77–90. doi:10.1007/s10551-013-1675-5.

Mikušová, M., and P. Horváthová. 2019. "Prepared for a Crisis? Basic Elements of Crisis Management in an Organisation." *Economic Research-Ekonomska Istraživanja* Vol. 32(1): 1844–1868. doi:10.1080/1331677X.2019.1640625.

Pfeffer, J. 2010. "Building Sustainable Organizations: The Human Factor." *Academy of Management Perspectives* Vol. 24(1): 34–45. doi:10.5465/AMP.2010.50304415.

PricewaterhouseCoopers. 2020. How COVID-19 Will Reset Traditional Workplace Rules. PricewaterhouseCoopers. https://pwc.blogs.com/the_people_agenda/2020/07/how-covid-19-will-reset-traditional-workplace-rules.html.

World Economic Forum. 2020. How the Post-COVID Workplace Will Change Business for the Better. World Economic Forum. https://www.weforum.org/agenda/2020/07/how-the-post-covid-workplace-will-change-business-for-the-better/.

About the Editors

Dr. Poornima Madan is an academician, researcher, and trainer with more than 14 years of corporate and academic experience. She is a PhD holder, UGC-NET qualified, and an MBA (HR) with two years of Diploma in Training & Development, ISTD. Presently, she is working as an Assistant Professor, Jaipuria Institute of Management, Noida, India. She has research papers published in ABDC-listed and Scopus-indexed journals of Emerald, Inderscience, and Sage. She is also the reviewer of various ABDC-listed journals. Her research interests are in the areas of employee engagement, mentoring relationships, personality, impression management, and managerial effectiveness.

Dr. Shruti Tripathi is Professor, School of Employability and Holistic Development, Delhi Skill and Entrepreneurship University, Delhi. She has her PhD in Quality of Life from the University of Allahabad, one of the oldest and top-ranked universities in India; She is NET qualified and has a first-class academic career. Her research interests include quality of life, sustainable development, quality of work-life, future of work, the emerging issues of management and technology like the impact of diversity, change in societal values on career decisions, etc. especially related to females. Her research contributions have been accepted in many international and national journals of repute. She is also an Advisor Member of the Planning and Monitoring Division of NCERT (National Council of Educational Research and Training), an autonomous organization of the Government of India. She is also the Editor for the book by Taylor & Francis bearing ISBN 9780367433994, titled *Quality of Life: An Interdisciplinary Perspective.*

Dr. Fehmina Khalique is a PhD and an MBA in human resource management and currently works as an Associate Professor, Head – Research Cell, Editor-in-Chief Lloyd Business Review and Lloyd Chronicles at Lloyd Business School, Greater Noida, India. She is associated with academics for the past 14 years in human resource management, business communication, and cross-cultural management. She has four years of industry experience and enjoys being a corporate trainer, counselor, and consultant. Her research interest lies in cross-cultural management, international management, human resource management, and human capital management, besides, she has several high-quality research papers to her credit. She is also involved in copywriting, editing, and reviewing research articles. In her free time, she likes to do some social work and deliver free career counselling sessions for students on honing their interpersonal and corporate-communication skills.

Ms. Geetika Puri is an academician and a trainer with 12 years of experience teaching Organizational Psychology, Human Resource Management, Compensation, and Labor Laws. Presently, she is associated with Gitarattan International Business School, New Delhi, and with K.K. Modi International Institute, Noida. She is also engaged with Philomath Research Pvt. Ltd. as a Director. Her research interests are in the fields of psychological contracts, education, organizational commitment, and personality.

Contributors

Gautam Bapat has been an Assistant Professor at MITSOM College in Pune since 2008 and is currently with MIT World Peace University's newly redesigned School of Management (UG). He is now appointed as the Head of School at the School of Management (UG). His knowledge in the fields of academia, industry, and consulting is impressive. He is the former Managing Director of ERiS InfoTech Pvt. Ltd., a firm that specializes in computer software development and consulting. He graduated from Fergusson College in Pune. He has earned a Master's degree in computer application (MCA – Science) and a Bachelor's Degree in Zoology. He went on to earn a Post Graduate Diploma in Business Management (PGDBM) with a Marketing Specialization after that. Dr. Gautam Bapat also placed first in his fourth degree, Master of Management Studies (MMS) with a Marketing Specialization. In the area of Strategic Digital Marketing, he has a PhD. He has extensive teaching expertise in the fields of computer science, applications, and management. He was also a member of the Ad-hoc Board of Studies for BBA (CA) at Savitribai Phule Pune University (SPPU) and the Board of Studies for MIT World Peace University. He is also a recipient of "Ideal Teacher Award 2018" by MAEER's MIT Group of Institutions. Dr. Bapat works in the field of social welfare as well. He is now Vice President of the Bapat Parivar Charitable Trust, an organization that works for the general welfare of society by providing school scholarships, sponsoring old age homes, and financially assisting poor individuals. He has authored 11 books on contemporary subjects like Computer Applications, E Commerce, and Digital Marketing.

Jimnee Deka is a Research Scholar at Amity University, Noida. She is drawing a scholarship from UGC under the UGC SRF grant. She has a bachelor's degree in Economics and has done her post-graduation with specialization in Financial markets. Her area of research interest is behavioural economics. She has to her credit conference presented paper, as well as publication in ABDC-indexed journal.

Mili Dutta is an Assistant Professor at Birla Institute of Technology-Extension Centre (BITEC) Lalpur, Ranchi, India. She is NET qualified, LLB, and BSc. She has more than 15 years of teaching and research experience. Her areas of interest are HR and Marketing. She has many publications of national and international fame to her credit like ABDC, Scopus, Web of Science, and UGC care.

Patrick Gibbons is Professor of Management at University College Dublin. His work has been published in outlets such as *Journal of Management Studies, Human Relations, British Journal of Management, Journal of Product Innovation Management*, and *Journal of International Management.*

Nimit Gupta is Professor of Marketing at School of Management & Liberal Studies – SOM – The NorthCap University, Gurugram, India. He is also CO-Head, NCRC, NCU Case Research Centre and Chair – NCU Alumni Committee. Dr. Nimit Gupta is the advisory board member of Harvard Business Review. He is a certified marketing trainer trained by Dr. Philip Kotler. He is certified from Harvard for case teaching. He has 19 years of experience in Teaching, Academic Administration, Consultancy and Research at reputed management institutes of Delhi NCR. He had worked in various capacities as PGP Chairperson, Programme Director, Internship Coordinator, NBA Team Member, Marketing Club Coordinator, Governing Board Member (Faculty Representative), Examination Controller to name a few. He has authored three books on contemporary themes – Permission Marketing and Case Writing. He has published 70+ research papers/cases/articles/book chapters in journals of repute including *Journal of Services Research, Journal of Electronic Marketing and Retailing, Journal of Digital and Social Media, Journal of Teaching and Case Studies, Journal of Marketing and Communication, International Journal of Customer Relations, Indian Journal of Marketing*, etc. He is a prolific case writer, and his cases are used for classroom teaching by academic institutions like University of Newcastle Business School-UK, Qatar University – Qatar, Bologna Business School – Italy, ICD International Business School – France to name a few. He is recipient of several scholastic performance awards including Best Case Award: Case Centre (USA), Dr. G.D Sardana Memorial Young Scholar Award –

George Mason University, USA & BIMTECH – India, Best Faculty Award, National Excellence Award for Innovative Teaching, World Education and Skill Conclave, Rastritya Srijan Siksha Rattan Samman to name a few. He is an Accredited Management Teacher from AIMA and has been instrumental in organizing MDPs in Marketing Management in association with Ministry of Micro, Small and Medium Enterprises, Government of India. He is on review and editorial board of national and international journals of repute like *Emerald Emerging Markets Case Studies* (EEMCS), *South Asian Journal of Marketing* (Emerald) to name a few.

Heikki Karjaluoto is a Professor of Marketing at the University of Jyväskylä School of Business and Economics, Finland. He is the leader of the Digital Marketing and Communication research group. His research interests include digital marketing, customer relationship management, marketing communications, mobile communications, and retail banking. He has published extensively in marketing and information system journals and is the author of several books. He has collaborated with several researchers both in Finland and abroad and actively cooperates with Finnish companies in joint research and development projects. He has won numerous research and teaching awards, including two Best Paper Awards from Industrial Marketing Management. He has published over 100 journal articles in journals such as the *European Journal of Marketing*, *Industrial Marketing Management*, *International Journal of Information Management*, *Internet Research*, *Journal of Business Research*, and *Journal of Travel Research*.

Navjot Kaur is a Managing Director of MoneyCom, Financial Services Company based in Queensland, Australia. She leads and assists the delivery of key transformation initiatives majorly in asset finance to optimize financial services and management models. She owns and directs key transformation deliverables including governance, resourcing, scheduling, risk management, change management, and stakeholder communication. She believes in promoting and enhancing the culture of the team through effective and open communication.

Being diversely qualified, she is also passionate about advocacy and social justice and serves as a Policy officer at Palliative Care Queensland. While performing a role with a difference, she undertakes research and analysis to identify emerging issues related to health, well-being, and equity for policy, planning, and advocacy purposes. She is responsible for responding to government inquiries, reviews, submissions, and reports to inform effective policy development. She has an MBA from Guru Nanak Dev University, Amritsar and Masters of Public Health from The University of Queensland. She has spent more than a decade as an educator, teaching courses related to Business Management, Emotional Intelligence and Business Sustainability in universities and registered training organizations in India and abroad. During the span of her career, she has also mentored and assisted diversity of investors to start up small- and medium-sized businesses across Australia.

Anuj Kumar is currently working as an Assistant Professor at the Apeejay School of Management, Dwarka, Delhi. He is pursuing an Executive-PhD in management from Aligarh Muslim University (Central University). He holds a double master's degree in management with a specialization in International Business. He has completed MSc in International Business from University College Dublin, Michael Smurfit Graduate Business School, Ireland. He has completed BTech from Guru Gobind Singh Indraprastha University, Delhi. He has more than 50 publications in Scopus and ABDC category journals. He also has three e-books and five book chapters to his credit (two are Scopus indexed). He is a Guest Editor of *Journal of Statistics and Management Systems* by Taylor & Francis.

Damien McLoughlin is Professor of Marketing at University College Dublin. His work has been published in outlets such as *Entrepreneurship Theory and Practice*, *Journal of Business Research*, *Industrial Marketing Management*, *Technovation*, and *European Journal of Marketing*.

Federica Pazzaglia is Professor of Management at University College Dublin. Her work has been published in outlets such as *Journal of Management*, *Journal of Management Studies*, *Human Relations*, *Human Resource Management*, and *MIT Sloan Management Review*.

Dipesh Ranjan is a Senior Vice President & Head-Asia Pacific & Japan, Mavenir at Singapore. He is an Official member – Forbes Technology Council. He is also a recognized leader in the telecommunications and cloud industry. He is renowned for his expertise in the new-generation technology ecosystem, as well as his ability to build businesses from the ground up while developing and managing high-performing, global teams. Prior to Mavenir, he led several key positions at Tata Communications before being named Vice President and Managing Director, Asia-Pacific, Japan and India, for NetFoundry, a Tata Group innovation business. There, he led successful initiatives with telecom operators, enterprises, and cloud providers, establishing various regional business units around enterprises and partners across India and Asia-Pacific and built a global partner group based out of the United Kingdom. He is an alumnus of the MIT Sloan School of Management, where he serves as MIT's education council representative in Singapore. Dipesh leads the APAC organization of Mavenir, turning vision into reality through sound strategy development and mentoring the team to be future leaders.

Ajaya Kumar Sahoo is the Founder and Director of Influer Ressource Humaine Solutions LLP, a New Delhi-based management and HR consulting firm. He is an MBA from the Faculty of Management Studies, Delhi University and a master's in political studies and international relations from JNU, Delhi. He spent 32 years in the corporate world before breaking out to become a full-time consultant-entrepreneur in 2016. During his corporate stint, he went on to garner rich experience in five sectors of the economy – power, automobiles, construction materials, infrastructure, and tyres, in India and abroad, encompassing MNCs, large and mid-sized companies like NTPC, Daewoo Motors, Eternit Everest, Lanco Infratech, Al Dobowi, and Zenises. In his last corporate assignment, Ajaya was the Global Group HR Head for two tyre majors, based out of Dubai and handled HR in the MENA region, South Africa and Europe and during which, he handled people operations for employees belonging to 29 different nationalities. Apart from consulting, he loves to teach. He is a guest faculty at IIM-Raipur, TISS (SVE), JBS, NSB, and ISTD. At JBS he teaches a course on cross-cultural management.

Meghna Sharma is a Professor of Economics & Strategy at Amity International Business School. She has 25+ years of experience in Industry and Academia. She has a PhD in Economics and has multiple research papers to her credit. Her research interest areas are Microfinance, Developmental Economics, Green Marketing and Sustainable Development, Strategy & Innovation. She has to her credit several publications in refereed journals, indexed in ABDC and Scopus.

Karan Sonpar is Professor of Management at University College Dublin. His work has been published in outlets such as *Journal of International Business Studies*, *Journal of Management*, *Journal of Management Studies*, *Human Relations*, and *MIT Sloan Management Review*.

Shruti Traymbak is an Associate Professor of HR and Marketing at Jagannath International Management School, New Delhi. She has more than six years of industrial experience and nine years of teaching and research experience. She has graduated from Miranda House, Delhi University and has been awarded PhD in Human Resource Management in 2019 from the reputed Institute Birla Institute of Technology, Mesra (Ranchi). Her areas of interest are HRM, Organizational Behaviour, Payroll Management, Marketing and HR Analytics, Training and Development and Industrial Relations and Labour Law. She has many publications of national and international fame to her credit like ABDC, Scopus, Web of Science, and UGC care.

Parul Wadhwa is an erudent professional who has worked with the CXO level for multiple Fortune 500 companies spanning across geographies. Currently she is associated with EXL Consulting, UK, as a Program Manager. She is a graduate from Delhi University with post-graduation from IMT Ghaziabad with industry certifications. She has lived and worked across 20 countries and currently based in UK London, where along with work enjoys hiking, sports like squash, tennis, etc., and exploring the culture through cuisine and conversations.

Shilpa Wadhwa is a PhD and MCom from Delhi School of Economics, BCom – H (Gargi College), University of Delhi and currently working as an Associate Professor with I Business Institute, Greater Noida. She has 14+ years of experience in academics and research. She is a certified trainer from The Indian Society of Training & Development and Behavior Assessment certification from Indraprastha University. Learning and sharing are her passion and has an interest in HR analytics and research. She enjoys guiding and counselling students in their career decisions. She has published research papers in national and international journals. She has reviewed books on business communication too. She enjoys calligraphy, sketching, and painting in her free time.

Leyla Yacine is a Doctoral Researcher on Emergent Work in the Digital Era (EWIDE) at the University of Jyväskylä, Finland. She is pursuing her postgraduate Business studies in Marketing. Leyla Yacine's doctoral dissertation working title is *Co-creating the future of work: leveraging algorithmic leadership for internal branding*. It is cross-disciplinary because it stands at the crossover between marketing, HRM, leadership, and information systems. She has authored another book chapter – "Video

Recruitment: Online Perspectives and Employer Branding Implications", to be released in Spring 2022 – besides the one published in this book. She believes organizations should consider their workers as "internal customers", whose needs and demands should be addressed effectively for them to be fulfiled and perform successfully. Besides her academic research, Leyla Yacine conducts freelance research covering the enthralling topics of employer branding, employee experience, Gen Z, Gen Y, and hybrid work, notably through The NextGen Project, a think tank based in Helsinki that aims at building dialogue across generations in organizations. In association with The NextGen Project, Leyla Yacine launched her well-being podcast, *The Inside Out Show*, in April 2022. She interviews talented professionals on building sustainable well-being practices in the digital era.

WORK

I

Chapter 1

Cross-Cultural Management, Anyone? Everyone!: CCM Is Going to Be Ubiquitous in the New Normal Post COVID-19 Pandemic and How Corporates Ought to Address It

Ajaya Kumar Sahoo

Influer Ressource Humaine Solutions LLP, New Delhi

Contents

DOI: 10.4324/9781003267751-2

1.1 Introduction

Yuval Noah Harari, in his seminal work *Sapiens*, tells us that 70,000 years ago, contemporaneous with the Cognitive Revolution* Homo sapiens moved from East Africa to other parts of the world. They did not have to bother about the nuances of the culture of the Neanderthals or Denisovans – they just moved in and exterminated them, over a period (Harari 2015). Neither did the invading armies of Alexander or the kings and emperors of the ancient and Middle Ages care much about the cultures of the peoples they vanquished and subjugated. They were marauders or occupying forces, out to extend the boundary of their empires, forcibly imposing their will over the conquered populace.

This was going to change in the 16th century, with the advent of the Joint-Stock Companies. The representatives of these companies came to the soon-to-be colonies as supplicants. In the early stages, these future masters and overlords had to learn the cultures and mores

* **Cognitive Revolution**: According to Harari, the "Cognitive Revolution" occurred 70,000 years ago causing the Homo sapiens' mind to shift, turning the species from "an insignificant African ape" into modern humans as "ruler of the world". Three things happened at that point of time in our history. Humans developed a language, acquired communications skills to interact with third parties and created, what Harari calls, "collective fictions or imagined realities" that gave humans the potential to bond and collaborate with virtually an unlimited number of members of their species.

of the then rulers and populace to be able to make inroads into these territories.

It is only with the advent of the international/multinational companies (MNCs) (those who did business across borders) that cross-cultural management (CCM) came into its own. It was a necessity rather than a good-to-have skill set. The CCM space attracted serious scholars and practitioners. People moved away from ethnocentrism[†] and self-reference criterion[‡] to try to understand the cultures of peoples around the world. Edward Hall, Florence Kluckholm & Fred Strodtbeck, Geert Hofstede and Fons Trompenaars, amongst others, have made seminal contributions to this field. Of course, there are "convergence of cultures" theorists (the world-is-flaters) who discount the study of CCM in the context of globalisation (Friedman 2007). But the very fact of globalisation underlines the need for, and continued relevance of, CCM like never before.

More so in the new normal of the pandemic which mandates work-from-home (WFH)/hybrid working for the foreseeable future. The new workplace/work practice has affected every aspect of human resource management (HRM) – be it recruitment and selection, onboarding, training and development, performance management, compensation and rewards, or employee relations. In the WFH era, recruitment has started to become geography agnostic (Business Process/Knowledge Process Outsourcing companies – BPOs/KPOs – started the trend a few years back, in a small way) and is likely to be more so with each passing year. And, once you have a multicultural workforce – no matter whether they are on your regular rolls or are gig employees/freelancers/contractual/consultants – all other sub-systems of HRM need to be rejigged to manage this eclectic workforce.

This chapter tries to look through the new paradigm looking glass as to how HRM needs to fundamentally reinvent itself with CCM as the *pièce de resistance*.

[†] **Ethnocentrism**: Evaluation of other cultures based on preconceived notions originating in the norms, values, standards, and customs of one's own culture.

[‡] **Self-Reference Criterion**: Self Reference Criterion is basically the frame of reference one uses while making decisions. Oftentimes unconscious, this frame of reference is coloured by one's own knowledge, experiences, and cultural values. In a sense, it tells the world at large that one's own culture is the best there is and other cultures need to be judged against this standard.

1.2 Organisational Culture, Cultural Distance,[§] Cross-Cultural Intelligence

Any country with reasonably large boundaries is bound to have a cornucopia of cultures and subcultures lending itself as an interesting subject of study. However, for this chapter, we have taken up only those corporates who do business across their national borders or who have employees (on regular rolls or otherwise) from other cultures/nationalities. Although a country of the size of, say, India, contains within its borders linguistic and cultural diversity of epic proportions and, as such, would warrant scholarly inquiry on its merit, this study restricts its scope to only cross-border transactions and the interface between employees from different nationalities in a corporate set-up.

But before that, we need to understand what we mean by organisational culture, cross-culture, and CCM. Prof. Edgar Schein of MIT's Sloan School of Management defined organisational culture as a band of basic assumptions held in common by the organisation members; and these shared basic assumptions develop over a period of time as the organisation grapples with the problems and challenges it faces, both internally and externally. The validity of these common beliefs develops over the years based on their ability to solve the problems the organisation faces (Organisational Culture and Leadership 1985). Geert Hofstede's seminal 1980 book, *Culture's Consequences: International Differences in Work-Related Values*, defines cultural distance as the extent to which the values and shared norms in a culture or country differ from those in another culture or country (Hofstede 2001). Cross-cultural intelligence or cultural quotient (CQ) is the ability to understand and work seamlessly and effectively across different cultures. This chapter will focus more on what culture does. What culture does is it predisposes a person to behave in certain ways, based on their upbringing, things they hold as self-evident in the process of coping with and making sense of the world they find themselves in. The situation gets further complicated when an employee must deal with people from other cultures. Individuals from the same country but different subcultures are comparatively easier to handle as the subcultures may overlap with some commonalities. But not so with alien cultures, if the cultural distance is too large, say, between China and a country in sub-Saharan Africa. Although

[§] **Cultural distance**: The extent to which the values and shared beliefs in a culture or country differ from those in another culture or country.

there may be some commonalities at the basic human level, then it is as far as it goes. To deal with intercultural differences, one needs CQ – having some knowledge of the national culture and mores. A disclaimer, though: one needs to use one's cultural knowledge as a heuristic rather than a deterministic device. But it is a start, a promising start, to begin to deal with a person from a different culture.

1.3 Cross-Cultural Skills in the New Normal

CCM was a must-have skill for some corporates beforehand too, for those who had anything to do with cross-border business. Even for purely domestic companies who have part of their supply chain across the border, or who export their produce or offer services to outside clients, a modicum of cross-cultural knowledge is essential. How does it differ now? In (and post) pandemic?

In the current, emerging, and future workplace, cross-cultural managers with intercultural competence are going to be ubiquitous. Earlier, only supply chain managers having foreign vendors or suppliers would need to have some cross-cultural skills. Similarly, those handling exports requiring interaction with foreign customers would bother about cross-cultural aspects of their interactions. It is another matter of whether they were formally trained on cross-cultural skills or acquired these on the go. In the new normal which drastically reduces the entry bar for employees from remote geographies to work for cross-border employers, remotely, almost every employee needs to be a cross-cultural manager.

So, let us look at the effect of geography-agnostic post-pandemic ecology on various sub-systems of HR.

1.4 Recruitment and Selection

Traditionally, some MNCs having subsidiaries in another country would prefer using the ethnocentric approach¶ (having expats from the parent country) for their overseas operations. Koreans, Japanese, and Scandinavians

¶ **Ethnocentric approach**: In the early stages of its entry into another country, an MNC may opt for an ethnocentric policy for staffing where it decides to fill key positions with parent country nationals rather than employing local staff.

want to do so. And these would be mostly for senior leadership positions/corner offices. While MNCs like HUL would use a polycentric approach** (opt for host country nationals) for these positions. During the 1990s post-liberalisation, newer airlines in India would typically hire expat CEOs as Indian talent in such an industry would be hard to come by. The IT industry even before the pandemic started hiring overseas talent at lower levels as well, especially in areas where domestic talent was lacking, viz. Artificial Intelligence (AI), Machine Learning (ML).

In the pandemic/post-pandemic era when remote/hybrid working has become the norm, the constraint of physically relocating the expats to the host/third country which used to be a daunting proposition is no longer valid. Employees can remain where they are, in their home country, and still work for their employer in a different country. Apart from enabling technologies, the near parity of compensation and benefits across countries also contributes to the trend of hiring expats across borders.

In January 2022 HCL Technologies made public its plan to hire 40,000–45,000 freshers for FY23 (Shinde 2022). HCL are, of course, no stranger to hiring from countries other than the host country (India). In fact, they have on their rolls employees from such diverse geographies as Romania, Hungary, Bulgaria, Guatemala, Costa Rica, Sri Lanka, Vietnam, Australia, New Zealand, Poland, Mexico, Ukraine, and Belarus. Globally HCL has around 7,000 overseas employees which they plan to ramp up significantly in the next four quarters. This, according to the CHRO of HCL, is because of the twin imperative of shortage of domestic talent as well as the need for diversifying organisational skill sets. For example, there are geographies like Ukraine and Belarus which, though contributing fewer numbers to the total employee strength, are important none the same as many of them are pursuing PhD which sets them apart for analytical work. With attrition touching 20% plus, other IT companies too have such ambitious hiring targets.

1.4.1 Implication for HR

Apart from training its staff in the talent acquisition (TA) function, HR needs to train the hiring managers as well as other interviewers in cross-cultural sensitivity. This sensitivity will pervade the entire process, starting from CV

** **Polycentric Approach**: On the other hand, an MNC may decide, from the very beginning, to manage the affairs of its foreign subsidiary with the host country's nationals. This is the polycentric approach to staffing.

screening right up to the final selection stage. One needs to understand that even such a basic document as a resume is culturally bound. While in a country like the US the candidates blow their own trumpet in their CV, often making a mountain out of a molehill, so to say, the candidates in the Scandinavian countries would be at the other end of the spectrum, what with the pervasive Janteloven unwritten social code of conduct (which discourages individuals from tomtoming their personal achievements and maintain a low profile in public) forcing the candidates to mask, tone down, and underestimate their achievements. The same holds during interviews as well. At the time of the interview, a candidate from the US, the UK, and even India may be outspoken and overreaching. On the contrary, a candidate from East Asia (say, China, Japan, Korea) may take a moment to answer a question. An untrained interviewer may see it as a sign of being unsure of oneself or inept, whereas it may be a mark of respect for the interviewer and the question being asked before giving a considered view. An interviewer looking for eye contact from a Japanese candidate may be disappointed. AI-based video interview software can potentially be culturally insensitive and discriminatory. For example, HireVue video-based interviews until recently provided a recruiter with a choice of non-verbal cues (captured during a video interview) like eye movements, body movements, facial expressions, clothing, voice nuances, etc. So, the TA and interviewers need cross-cultural training before hiring cross-border candidates.

1.5 Compensation and Benefits

After selecting a candidate and before making an offer, HR must tread the minefield of setting and negotiating a compensation package with caution. The majority of employee dissatisfaction stems from an ill-conceived compensation and benefits package leading to the employee feeling cheated. The compensation offered must agree with the twin criteria of internal and external equity – employees having the same qualification and experience and doing similar work must be compensated equally; also, the compensation should be broadly comparable with the packages on offer in the same industry for the same or similar roles. If internal equity is not maintained and the new employee has been offered salaries much above similarly placed existing employees, the organisation runs the risk of losing such aggrieved employees. The reverse may push the new employee out.

When overseas candidates come into the equation, a few more variables conspire to queer the pitch. Variables like the differing cost of living, the standard of living, different income tax rates, social security benefits, etc. need to be considered while fixing the compensation of an overseas candidate. In this regard, organisations hiring expats normally follow one of the two methods for fixing compensation:

1. The Balance Sheet Approach or Capital Maintenance Approach
2. The Going Rate (or Market Rate) Approach

The Balance Sheet approach seeks to achieve purchase power parity between the home country and host country of the employee. Normally four buckets of usual household expenses are compared between the home and host countries, viz. income and other statutory taxes/deductions, residential accommodation and utilities, goods and services, and discretionary expenses. The employee is compensated for the difference and certain additional compensation (hardship allowance) is given for posting in less developed countries.

In the Going Rate (Market) approach, the salary of every employee – host country and overseas alike – is linked to the salary structures in the host country. The company, of course, exercises its discretion as to which salary structure it'd choose from amongst the options available in its industry and at what percentile it wants to position itself.

1.5.1 Implication for HR

The main purpose of compensation in an organisation is to influence workforce behaviour which, in turn, affects organisational performance. The primary objective of an effective compensation design is to ensure equity and efficiency. Equity is ensured at three levels – internal equity, external equity, and individual equity. Internal equity refers to similar jobs getting paid similar compensation, and more difficult jobs getting paid more. External equity ensures that jobs are fairly compensated in comparison to similar jobs in the labour market. Individual equity ensures equal pay for equal work. The efficiency function of compensation mandates that apart from linking compensation to productivity/profit/individual performance, it should also attract, reward, motivate, and retain highly capable and efficient talent. It should help the organisation maintain market competitiveness, ensure compliance with laws of the land, and help build the employer brand. When designing compensation in the new normal, HR needs to do a tight rope walk not only between internal and external equity but also between the home country and host country. For this, they need to improve their knowledge of the laws governing compensation in both countries, salary surveys, cost of living, etc. Apart from negotiating with the candidate, it also has the additional task of explaining the difference in compensation to the existing employees, should such an occasion arise.

1.6 Onboarding

Pre-onboarding phase (starting from the candidate accepting the offer till the first day of joining) is much more important in the case of overseas candidates than it is for domestic offers. It is important to keep candidates warm till they join. It is customary for organisations to introduce new joiners to their team with whom they'd be working, going forward. In the cross-cultural context, the organisation needs to broaden the group of employees much, much larger to introduce to the new employee. And, because of the online nature of the orientation/induction programme, it does not matter how big the group is. It is familiarity with people that ensures the new employee eases into the organisation much better. It is also important that regular check-ins with the new employee happen without fail, either through chatbots or personally followed by one-on-ones in case the check-ins throw up any flight risks through sentiment analysis.

1.6.1 Implication for HR

In many organisations, onboarding is either given a go by and employees are put on the job straight away or the function does not get the importance it deserves. In the process, the new joiner may feel alienated causing flight risk or they are acculturated in a dysfunctional manner leading to future behaviour issues. HR needs to ensure that the orientation/induction/onboarding process is robust and regular check-ins happen with the new overseas joiner. Cultural sensitisation (on either side – the new joiners as well as existing employees/managers) should be an integral part of the orientation process. It may help to provide the new joiner with formal mentoring support, at least for the first six to nine months. The organisation can try the "buddy system" pairing up an overseas employee with an existing employee from the host country.

1.7 Performance Management

"That which is measured improves. That which is measured and reported improves exponentially" (variously attributed to Karl Pearson or Peter Drucker). The importance of performance management in organisations can never be overemphasised. But employees, HR, and managers have always had a love-hate relationship with the performance management system. Bock (2015) in his book Work Rules titled one of the chapters "Why Everyone Hates Performance Management".

If performance management systems (PMSs) are complicated in the domestic environment, it is doubly so when overseas employees and managers are involved, coming from different cultural backgrounds. Performance management typically involves setting individual goals aligned to organisational goals, appraising, and measuring those goals periodically/ at year-end, and giving feedback apart from using the appraisals for various purposes like giving increments, promotions, development, etc. Culture affects every stage of performance management. During the goal-setting stage, power distance[††] (high/low), locus of control[‡‡] (inner-external), and poly/mono-chronic cultures tend to affect the process. In highly individualistic cultures like the US, it is easier to set mutually agreed goals through discussion between the employee and supervisor. But in high power distance cultures like that in India, it is the manager or supervisor who sets the goals for the subordinate employee. Even left to themselves, employees with an external locus of control and polychronic orientation may not be able to assert themselves enough to negotiate mutually agreed goals. When it comes to appraisal, the organisation needs to decide who should be the appraiser for an overseas employee or an employee located in a foreign country. Whether it should be a parent country manager or a host country manager or a combination of the two. While different geographic locations can be an issue, what muddies the water further is if the employee and the appraising manager come from different cultures. Giving and receiving

---

[††] **Power Distance**: In every group, be it a family, society, or organisation, power is unevenly distributed, some being superior, and others subordinate. In high power distance countries, this is accepted whereas, in low power distance cultures, this is resisted by those lower in the pecking order.

[‡‡] **Locus of Control**: Who controls your destiny? Who decides whether you should be happy or not? Are you being controlled by powerful others or fate or even God? If so, you have an External Locus of Control. People with an internal locus of control don't go looking for others to blame for what happens to them. They simply take control.

feedback, too, gets affected by different cultures. In high context[§§] cultures many things remain unsaid, but understood, whereas, in low context cultures, one needs to spell out everything in so many words. In far eastern cultures like China, there is a concept known as saving face (mianzi). An employee shall not say anything due to which a manager or peer may lose face. Similarly, a subordinate shall not take kindly to something negative spoken directly to her. So, feedback becomes a virtual tightrope walk.

1.7.1 Implication for HR

One size fits all PMSs in the new normal will not work. HR needs to have a close, hard look at the legacy PMS process in the post-pandemic ecosystem having overseas employees either working remotely or offline. And make necessary changes in the process. The HR needs to train themselves as well as other managers handling overseas employees in the cultural nuances of such employees they will be managing. Managers and supervisors need to be aware of concepts like power distance, locus of control, mono/poly-chronic[¶¶] orientation, high/low-context cultures, "saving face", etc. which affect almost all stages of the PMS process. It is not about giving up or suppressing their own culture but being aware of other cultures so that misunderstandings do not happen.

1.8 Training and Development

The training and development function will not be as straightforward in an online ecosystem as it has been previously. In the pre-pandemic era with the majority of employees operating from offices or factories and of uniform nationality, the training process would mean training needs identification/ analysis, training calendar, training design, implementation, and evaluation in a straightforward linear manner in a training year. The new normal comes

---

§§ **High-Context/Low-Context Cultures**: In high-context cultures, people leave out a lot in communication and expect the receiver to fill in the details through contextual assumptions. And neither the sender nor the receiver thinks much about it. They take it for granted. While in low-context cultures, everything must be spelt out, in black and white, in so many words. High-context cultures are collectivist; they value interpersonal relationships and have members who form stable and close relationships.

¶¶ **Poly/mono-chronic cultures**: People in monochronic cultures take up one thing at a time to do and see it through. They are orderly and punctual, sticking to agreed timelines. While people in polychronic cultures multi-task, they may not honour appointments on time or stick to agreed completion schedules for a project.

with its decisional problems. First and foremost, *who* to train? Will it be only the employees on the regular rolls of the company or even the gig employees, freelancers, or contractual as well? The organisations would ignore the training needs of their contingent workforce at their peril. Since the training, going forward will be almost all-pervasive, the line of sight linking training to business results will be imperative to justify every training dollar. Training needs to be identified at all three levels – organisational level, functional level, and individual level. And this need for identification must happen throughout the year. Given the short shelf life of many skills, training must be much more agile than ever before. The need to reskill and upskill large chunks of the employees will be the order of the day. Training Calendars will be a relic of the past because such calendars will become obsolete even before the ink is dry on them. Training needs to be on tap 24/7/365. The responsibility for learning will squarely rest on the employee, the Chief Learning Officer (CLO) playing the role of a facilitator, and curator of learning resources.

1.8.1 Implication for HR

Training function in the new normal will require a major mindset change not only on the part of HR but also for the employees and their managers. HR needs to learn to let go. The employees need to shift their locus of control to be more and more internally driven. The who, what, when, why, and where of the training too will undergo a sea change – being employee-centric and business result focused. Culture from where an employee comes from will affect the training process at every stage. Employees from individualistic cultures would want to have a say in which training programmes they are sponsored for. And also, which skills they'd want to learn and hone. Employees from collectivist cultures may not give genuine feedback on the training programmes they attend to "save face". Employees from low power distance cultures are more likely to question the credibility and authority of the trainer/facilitator and the latter needs to tweak the training methodology from trainer-led to trainee centric.

1.9 Employee Engagement

Employee engagement would need a paradigm shift in the new normal. Instead of it being an eat-and-drink-as-much-as-you-can kind of weekend "activity", it needs to take centre stage. Moreover, a mere rehash of physical engagement activities to suit the online medium is not likely to work. It can be counterproductive. HR needs to reimagine employee engagement. For the new employees, especially the overseas employees, employee engagement needs to start much before the candidate becomes an employee and needs to continue for a reasonable length of time, say, 6–9 months. New employee check-ins have to be more frequent, with the help of AI chatbots, followed by sentiment analysis and one-on-ones for possible flight risks. For older employees, periodic performance appraisal check-ins need to double up as employee engagement interventions.

1.9.1 Implication for HR

Employee engagement would warrant a paradigm shift in perspective in the new normal. No longer will so-called weekend "engagement activities" suffice to keep the employees engaged. The engagement shall have to be an employee life cycle process. Employee engagement shall be measured by touchpoints – how often the organisation reaches out and touches its employees' lives in increasingly innovative and creative ways. Also, cutting edge HR technology like wearables, chatbots, and Organisation Network Analysis will come to the fore, relegating traditional employee engagement metrics like Q12 to the background.

1.10 Communication

The basics of communication are the same world over – you have a sender and a receiver; the sender encodes the message, chooses a channel to carry the message, and the receiver decodes the message and sends feedback. Simple. Although it does look like everyone is dancing to the beat of the same drum, one finds to one's chagrin that not all are playing by the same set of rules. As long as employees/managers communicate within the same culture, things are under control to some extent. Bring in different cultures into the equation and the maths does not add up. Culture affects every stage of the communication process, be it encoding, choice of channel/ medium, decoding, or feedback. It even adds to the "noise" within which communication takes place, obfuscating the meaning that the receiver may glean out of a message. In the encoding stage, an employee from a high-context culture may leave out a lot of information presuming these to be understood which, to an employee from a low context culture, will be incomprehensible. Even knowledge of and facility with the language of communication will have an effect during encoding. The same cultural dimensions will affect the decoding of the message at the other end as well. Culture may affect the meaning the receivers give to the message received, which may be different from the meaning the sender gave to it at the time of encoding. In high power distance cultures, feedback may not happen or, if it happens at all, it may not be authentic. Not only verbal communication but even non-verbal communications are also fraught with cultural sinkholes as body language differs from culture to culture and may take on an entirely different (and sometimes unpleasant) meaning altogether than what was originally intended. And, additionally, there is the issue of not being privy to body language in online communication.

1.10.1 Implication for HR

Soft skills like communication, personal and interpersonal effectiveness, etc. are not taught in schools and colleges. Research shows that 85% of success in the corporate world is dependent on soft skills and only 15% on hard skills like engineering and IT knowledge. Some organisations put their new hires through campus-to-corporate programmes. But many don't. In the new normal, however, not only recruits, but even the existing employees need training on cross-cultural communication. The organisation needs to train its employees in differing cultural dimensions, including non-verbal communication, so that misunderstandings do not happen during cross-cultural communication.

1.11 Negotiation

Certain dimensions of culture may affect negotiations in a cross-cultural context. For example, employees belonging to task/deal-focused cultures like the US may be more interested in closing the deal and signing the agreement first before developing any sort of relationship. Whereas in a relationship-focused culture like Japan and China, this process is upended – business deals *follow* relationship-building instead of the other way around. Again, in specific cultures like the US and UK, interpersonal relationships are strictly official and professional. In diffuse cultures like China or the Arab countries, the boundaries between professional and personal are frequently blurred. In high-context cultures, the agreements may be vaguely worded; whereas in low-context cultures a lot of emphasis is laid on every word of an Agreement which remains sacrosanct throughout the business relationship between two entities.

1.11.1 Implication for HR

Organisations need to raise the awareness and sensitivity of their employees in respect of these cultural nuances in negotiation, viz. relationship vs task/deal focused cultures, specific vs diffuse cultures, high-context vs low-context cultures. In cultures like the US and some western and westernised Asian countries, contracts and agreements are sacrosanct and need to be adhered to in letter and spirit whereas some countries of the world are lax. So, one must be aware of these cultural mores before entering into a deal. Also, there are some culture-specific etiquettes regarding the exchange of business cards, gifts, table manners, etc. so that one may be able to avoid unintended *faux pas*.

1.12 Conclusion

McKinsey & Company in an article published in May 2020 and titled "To emerge stronger from the COVID-19 crisis, companies should start reskilling their workforces now" stated that adapting employees' skills to the post-pandemic ways of working will be crucial to managing and thriving in the new ecosystem (McKinsey 2020). The article suggests six steps to reskilling. The first two steps are:

1. Rapidly identify the skills your recovery business model depends on
2. Build employee skills critical to your new business model

HR world over has a very crucial role to play in the post-pandemic ecosystem. Especially from the standpoint of cross-cultural fluency. The new normal will affect and change each sub-system of HR more fundamentally like never before. As we have seen above, not a single sub-system will stay untouched by culture. So, not only does HR have to train their own, but also they need to train almost every other employee in their organisation in various aspects of CCM.

Before the pandemic organisations operating exclusively or predominantly in a domestic context could safely ignore cultural aspects of business and still succeed. Even corporates with some exposure to overseas clients/vendors, etc., say, exporters or importers, could do with minimal cultural training for a very small section of their employees. But no more in the new normal which makes it easier for organisations to hire overseas employees who may work remotely, possibly permanently. Hence the crying need to broad base cultural training in the organisation, for the existing and new employees, for all nationalities. So that they can effectively interact with one another without much meaning lost in translation.

References

Bock, Laszlo. 2015. Work Rules: Insights from Inside Google That Will Transform How You Live and Lead. London: John Murray.

Friedman, Thomas L. 2007. The World Is Flat: A Brief History of the Twenty-First Century. New York: Picador.

Harari, Yuval Noah. 2015. Sapiens: A Brief History of Humankind. London: Vintage.

Hofstede, Geert. 2001. Culture's consequences: Comparing Values, Behaviors, Institutions, and Organizations across Nations. Thousand Oaks: Sage Publications.

McKinsey. 2020. "To emerge stronger from the COVID-19 crisis, companies should start reskilling their workforces now". Agarwal, Sapana, Aaron De Smet, Sébastien Lacroix, and Angelik. Article. Accessed on 03.04.2022. https://www.mckinsey.com/business-functions/people-and-organizational-performance/our-insights/to-emerge-stronger-from-the-covid-19-crisis-companies-should-start-reskilling-their-workforces-now.

Shinde, Shivani. January 18, 2022. "HCL Tech to hire 40,000-45,000 freshers for FY23 to meet attrition threat". *Business Standard.* Accessed on 03.04.2022. https://www.business-standard.com/article/companies/hcl-tech-to-hire-40-000-45-000-freshers-for-fy23-to-meet-attrition-threat-122011701042_1.html.

Further Readings

Adamovic, Mladen. 2022 "How Does Employee Cultural Background Influence the Effects of Telework on Job Stress? The Roles of Power Distance, Individualism, and Beliefs About Telework". Research Article. *International Journal of Information Management.* Vol 62, 102437. https://doi.org/10.1016/j.ijinfomgt.2021.102437.

Festing, Marion. July 24, 2020. "Cross-Cultural Virtual Teams Are on the Rise, but Can They Communicate Effectively?" Accessed on 03.04.2022. https://blogs.lse.ac.uk/businessreview/2020/07/24/cross-cultural-virtual-teams-are-on-the-rise-but-can-they-communicate-effectively/#:~:text=By%

20embracing%20its%20diversity%2C%20cross,decision%20making%20and%20idea%2Dgeneration.&text=To%20structure%20an%20effective%20meeting,meeting's%20agenda%20and%20goals%20beforehand.

Hyun, Jane, and Douglas Conant. 2019. "3 Ways to Improve Your Cultural Fluency". *Harvard Business Review.*

Lockwood, Nancy R and Steve Williams. Third Quarter 2008. "Selected Cross-Cultural Factors in Human Resource Management". *SHRM Research Quarterly.*

Logemann, Robert. Jaunary 2021. "Remote Work Might Force Us to Be Better Cross-Cultural Managers". *Forbes.*

Mangla, Namita. 2021 "Working in a Pandemic and Post-Pandemic Period: Cultural Intelligence in the Key". *International Journal of Cross-Cultural Management.* Vol 21(1): 53–69. https://doi.org/10.1177%2F14705958211002877.

Roux, Peter W. 2021. "Developing Cultural Intelligence (CQ) in Blended Environments: Towards an Appraisal of Experiential Learning". *International Journal of Educational Media and Technology.* Vol 15(1): 105–116.

Schein, Edgar, Organizational Culture and Leadership (1985). University of Illinois at Urbana-Champaign's Academy for Entrepreneurial Leadership Historical Research Reference in Entrepreneurship, Available at SSRN: https://ssrn.com/abstract=1496184

"Working in Multi-Cultural Teams: A Case Study". 2022. *Deloitte.* Accessed on 03.04.2022. https://www2.deloitte.com/au/en/pages/human-capital/articles/working-multicultural-teams.html.

Chapter 2

Hybrid Work: Gen Z Expectations and Internal Employer Branding Implications

Leyla Yacine and Heikki Karjaluoto
University of Jyväskylä, Finland

Contents

DOI: 10.4324/9781003267751-3

2.1 Introduction

Gen Z – also known as "Generation Z" or "iGeneration" born between 1997 and 2009 – has been hit the hardest by the pandemic and remote work. In contrast to other generations, Gen Z has been struggling the most in terms of engagement or excitement about their work (Microsoft 2021), considering that they have been denied the opportunity to participate in proper onboarding, networking, and training that could help them better integrate within their organisation and team (WEF 2021). Gen Z's engagement at work already translated into a greater focus on workplace well-being before the pandemic and has now been amplified (McKinsey 2021). As the global workforce will include nearly 27% of Gen Zs by 2025, addressing their needs and preferences seems essential. In particular, this young generation is utterly compelled to work both on-site and remotely (Mărginean 2021; Pataki-Bittó and Kapusy 2021) and it intends to merge work and life (Barhate and Dirani 2022, 16), although most of them feel dissatisfied with their work-life balance (56%) and their jobs overall (59%) according to an Adobe Survey run in 2021. Hence, their well-being could be favoured through hybrid work, a new work arrangement emerging for white-collar workers across organisations around the globe.

Hybrid work is a mix of work performed in the office and remote work conducted from any place that provides internet access. It connects people across physical and virtual spaces with implications for work configuration and management (Halford 2005, 19). Hybrid work requires flexibility and coordination (Lin 2021, 1), but most importantly it builds on trust (Julka 2021, 6) in times of instability (Driscoll 2021, 22). This unstable context has exacerbated Gen Z's tendency to switch jobs and organisations if dissatisfied with their professional situation (Jayathilake et al. 2021). Motivating, engaging, and developing Gen Z as white-collar workers is therefore paramount amid the "Great Resignation" (Barhate and Dirani 2021, 15). This endeavour can be supported through internal employer branding, that is branding intended for the existing organisational talent.

Gen Z has been extensively researched in terms of career aspirations (Barhate and Dirani 2021), attitudes, beliefs, and expectations when entering the workplace (Gabrielova and Buchko 2021; Mărginean 2021), generational characteristics, attraction and their organisational implications (Pandita 2021; Pichler et al. 2021; Hernandez 2020; Mahmoud 2021; Schroth 2019), person-organisation fit (Graczyk-Kucharska and Erickson 2020), employee development and retention (Jayathilake et al. 2021; Schroth 2019). Still, extant studies have mostly involved Gen Z students whereas research with Gen Z workers remains relatively scarce (Gabrielova and Buchko 2021; Jayathilake et al. 2021).

Parallelly, hybrid work is currently receiving growing interest from researchers in the middle of the pandemic, especially when it comes to capturing what it is and navigating it (Babapour Chafi et al. 2021; Britton 2021; Driscoll 2021; Hepburn et al. 2021; Poggi 2021), its spatial and design characteristics and implications (Lahti and Nenonen 2021; Strategic Direction 2021; Halford 2005), trust (Julka 2021), or specific industries (Deckert 2021). However, there is not any research specifically targeted at the study of Gen Z white-collar workers concerning hybrid work.

Additionally, there is a lack of research on internal employer branding amid the pandemic with very little work shedding light on employer branding in general (Nelke 2021) or external employer branding to attract Gen Z (Pandita 2021; Pichler et al. 2021).

This chapter aims to fill the gap in extant research by combining the concepts of Gen Z, hybrid work, and internal employer branding to find how organisations can motivate and engage Gen Z white-collar workers in the context of hybrid work. To be more precise, this chapter explores Gen Z's viewpoints of hybrid work and advances recommendations to enhance internal employer branding with this young generation, ultimately compelling its members to stay. In this regard, it responds to the call for professionalisation and adaptation of post-pandemic internal employer branding (Nelke 2021) by probing Gen Z, who directly experience the employee value proposition (Pandita 2021, 11). The study of Gen Z expectations in the labour market should help organisations re-envision, adapt, and refine their employee value proposition to ensure Gen Z workers' well-being, engagement, satisfaction, and retention. To drive the change, this chapter first conflates business, HRM, marketing, social, and information technology literature to develop a theoretical framework that drives the study of Gen Z white-collar workers regarding what they expect from their employers in hybrid work arrangements and how internal employer branding can be improved accordingly. Next, the methodology is elaborated on before presenting the research findings and their concrete implications for (1) organisation and leadership, (2) Gen Z workers, and (3) social cohesion and teamwork. Finally, the chapter closes with future research avenues on Gen Z, internal employer branding, and hybrid work.

2.2 Literature Review and Theoretical Framework

This section presents and correlates the three core tenets of this chapter, namely Gen Z, employer branding, and hybrid work. Gen Z characteristics, expectations, and their effects on leadership and organisational practices can serve for employer branding in the context of hybrid work, as crystallised in the theoretical framework introduced at the end of this section.

2.2.1 Gen Z Profile

Gen Z constitutes the youngest working cohort that will soon rise in the global workforce. They were raised in a world permeated by technological advancements that enable uninterrupted access to a gigantic ocean of information, where anxiety and isolation easily arise and develop (Hernandez 2020). The youngest population segment of working age, therefore, has a knack for all things technology (Barhate and Dirani 2021; Hernandez 2020; Mahmood et al. 2020; Schroth 2019). In the digital era, Gen Zs especially spend their time on social media platforms where anyone can express themselves, create, share, react to, and comment on content instantly, thereby diluting spatiotemporal boundaries between people as well as entities across the world (Hernandez 2020). This implies that they can access any type of information online, with the potential to build or erode any entity's (individual, brand, or organisation) reputation through content virality (Hernandez 2020). Nevertheless, constant technological connection and activity may contribute to a state of anxiety (Hernandez 2020; Schroth 2019), isolation and undermined well-being (Mărginean 2021; CIGNA 2018; Weller 2017) that can lead to depression (Pichler et al. 2021).

Interestingly, Gen Z's active use of social media does not seem to preclude them from preferring face-to-face (F2F) communications over online ones in their quest for authentic relationships (Mărginean 2021; Half 2015; Tulgan 2013), motivation (Pataki-Bittó and Kapusy 2021), and person-organisation fit (Graczyk-Kucharska and Erickson 2020). Gen Zs want to fit into their organisation and team, in which they could build strong relationships (Barhate and Dirani 2021) and express themselves freely and genuinely (Gabrielova and Buchko 2021). Besides, "authenticity reinforces trust", one that is first established through branding in the online realm (Hernandez 2020, 20).

When it comes to career and personal development, the youngest and most educated workers (Schroth 2019) not only want to feel passionate about their work and benefit from clear guidance and mentorship, but they also are more prone to join employers that offer flexible, informal, and more relaxed environments (Gabrielova and Buchko 2021; Mărginean 2021; Graczyk-Kucharska and Erickson 2020). They are motivated by learning opportunities to develop themselves laterally within their organisation, they need clear instructions provided by meaningful mentors to grow successfully on their learning journey (Barhate and Dirani 2022; Schroth 2019). On this journey, they want autonomy to experiment, create, innovate, make mistakes, and learn from them (Pandita 2021; Pichler et al. 2021), which thereby reinforces the high value they grant to feedback (Gabrielova and Buchko 2021; Turner 2015) and recognition (Pataki-Bittó and Kapusy 2021; Pichler et al. 2021). This need for autonomy tied to Gen Z's online activity and their need for inspiration could explain their preference for self-reliant learning methods, amongst which social media, web searches (Pichler et al. 2021), and YouTube videos (Pearson Higher Education 2018) prevail.

Beyond that, Gen Zs are seeking the flexibility and adaptability that they believe are favourable to their desired lifestyle and successful workforce business drivers (Deloitte, 2021) whilst they have grown risk-averse and financially conscious (Koop, 2021). Especially flexible working hours, stability, and financial security are expected from employers (Barhate and Dirani 2021; Pataki-Bittó and Kapusy 2021) to reduce their anxiety and, therefore, enhance their well-being (Hernandez 2020). To Gen Z, work accounts for a major and "important part of their life", thereby amplifying their hedonistic preferences, such as enjoyment and entertainment, when it comes to employment choices (Pataki-Bittó and Kapusy 2021, 165).

Gen Zs long for fair, transparent (Graczyk-Kucharska and Erickson 2020), diverse, and inclusive working environments, where they could find a greater sense of commitment than older generations (Mahmoud et al. 2021). They comprise the "most diverse generation ever" and display higher tolerance for others' different beliefs (Pichler et al. 2021; Schroth 2019). Growing up in an era of social justice movements contributed to shaping their thirst for and sense of fairness (Schroth 2019) and consistency (Hernandez 2020).

Now that Gen Zs are joining professional teams and disrupting both the latter and organisational ways of working, they should be studied from their new lens of workers rather than students, unlike past research (Barhate and Dirani 2022; Gabrielova and Buchko 2021; Graczyk-Kucharska and Erickson 2020; Hernandez 2020). Specifically, Gen Z's workplace behaviours may not be foreseeable from data collected with students and distinct theories should be developed in an attempt "to incorporate Gen Z's characteristics unique to their generation" (Barhate and Dirani 2021, 15). In this regard, understanding Gen Z workplace drivers and motivators has become crucial for HRM and organisations to create working environments of value where the youngest talent can thrive (Mahmoud et al. 2021). One such theory that facilitates Gen Z understanding can be recognised in the work values coined by Pataki-Bittó and Kapusy (2021) and presented in Figure 2.1.

The authors identified five key work value categories in Hungarian Gen Z employees looking for a job before the COVID-19 pandemic, namely: extrinsic values, intrinsic values, growth/power values, social values, and convenience values. Extrinsic values "provide tangible rewards" that allow for stability and security and include compensation, high wage, bonuses, organisational services, and other perks and benefits. Intrinsic values refer to essential values from which mental rewards spur, such as the meaningfulness of work and organisational purpose, enjoyment, autonomy, challenges, creativity, interesting work, corporate social responsibility, ideological fit, recognition, and the like. Growth/power values are for the benefit of individual development and status through and at work. Social values feed the need of social connection at work and comprise team building, team spirit, teamwork, positive impression of coworkers and managers, and so on. Convenience values relate to health support, support in the office, and work-life balance support. Although this Gen Z work values model provides a comprehensive talent attraction overview in the light of job seekers' values, it was recommended to test it against the work values of Gen Z workers who are employed in the context of the pandemic (Pataki-Bittó and Kapusy 2021).

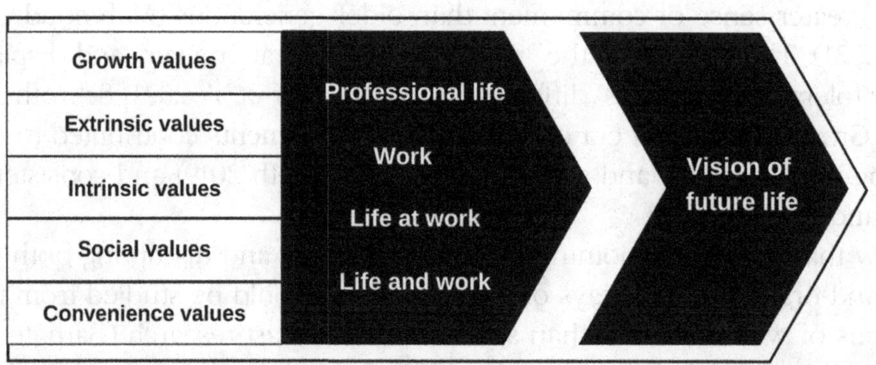

Figure 2.1 Gen Z work values. (Modified from Pataki-Bittó and Kapusy 2021.)

Ironically, the iGeneration workers have faced major difficulties amid the COVID-19 pandemic that granted them the flexibility they crave: according to an Adobe Survey, they have struggled more than other working generations in maintaining a work-life balance in remote work arrangements and have felt pressured to cling to office working hours, that is 9:00–17:00, while they may perform at their best outside this time window. Those who started a new position or were in their early career stages during the pandemic have seen their socialisation process along with learning and development detrimentally affected because they missed out on key cultural elements (informal cues and exchanges for instance) and phenomena (events, fairs, physical coworking practices, etc.) due to lack of in-person interactions of working from home (WFH, a type of remote work model) schemes (Nelke 2021). This was also stressed by Microsoft (2021), who: (1) found that the increase of asynchronous communications at the expense of synchronous communications results in difficulties "to exchange more complex information, converge on the meaning of information, and build relationships", (2) warned against the decrease of access to new information and innovation loss related to a distancing from weak social ties in comparison to the strengthening of strong ties, and (3) purported that social connections can be tapped into overtime under remote work circumstances, although the odds are higher for connections built-in person (Babapour Chafi et al. 2021, 2).

2.2.2 Internal Branding Targeted to Gen Z in New Working Models

Hence, organisations should review their ways of working and management to possibly cater to Gen Z's needs and preferences and ensure their integration, engagement, and retention in the context of new working arrangements such as WFH and hybrid work, which are both set to last (McKinsey 2021; OECD 2020). This entails an upgrade of internal employer branding practices to become or remain a good (if not great) place to work. "Internal employer branding", or "internal branding", refers to employer branding directed towards the existing workforce in contrast to "external (employer) branding", which is aimed at the attraction of new workers (Yacine 2021). "Internal branding is a process that aims at the creation and infusion of a corporate culture of trust between the employer and their employees, or establishing strong corporate moral values, from which can spur employees' pride, or satisfying employees through the fulfilment of their psychological contracts" (Sengupta et al. 2015, 308–309; Yacine 2021, vi). Employer branding builds a distinctive brand as compared to other employers on the labour market and should grant a competitive advantage to organisations that are and will be able to renew themselves, their culture, and practices during and beyond the pandemic (Nelke 2021; Pandita 2021). Pandita (2021) posits that employer branding endeavours and practices that create and support diversity, an innovation mindset, a collaborative culture, a flexible environment together with the community will attract Gen Z workers. Similarly, Pichler et al. (2021) propose that employers that follow the DITTO framework, including diversity, a balance between individualism and teamwork, technology, and organisational support, will entice Gen Z workers. In other words, the employer brand should be targeted to Gen Z to appeal to them by aligning with their values, characteristics, needs, and preferences that further tailor the employee value proposition (i.e., the employer offer, similarly to what the value proposition is to customers) to them in new working models (Nelke 2021). Considering employer branding experts' responses to a survey obtained in July 2020, Nelke (2021) pinpointed "the urgent necessity for companies to professionalise their external and internal employer branding via online channels" while planning employer branding across online and offline tools (392). This converges with the rising adaptation required for hybrid work.

2.2.3 Hybrid Work: Multi-Located and Connected Work

Amongst the new working arrangements, hybrid work has emerged as a solution to the challenges of in-person work and those of remote work, thereby offering the advantages of both to white-collar workers. Hybrid work refers to work undertaken partly on-site on the employer's premises by some white-collar workers and partly remote by some other white-collar workers, and the "when" and "where" of work may freely be decided by white-collar workers (Babapour Chafi et al. 2021) who are all connected in the online realm through devices and apps (Neumayr et al. 2018). Hybrid work is multi-located (Halford 2005) and implicates hybrid collaborations and hybrid meetings. Hybrid collaborations consist of "collaborative practices that involve simultaneous co-located and remote collaboration with phases of both synchronous and asynchronous work that spans multiple groupware applications and devices" (Neumayr et al. 2018, 1). Hybrid meetings associate on-site and remote participants (Hirsch 2021; Neumayr et al. 2021). "Hybrid work" envelopes different types of work arrangements across people, places, and spaces; a worker can experience a few of them with one employer and the latter may resort to different hybrid arrangements for different workers in their ranks (Lin 2021). Moreover, hybrid work appeals to white-collar workers, who majorly want to adopt it (Hirsch 2021; McKinsey 2021).

Hybrid work has been underexplored and its multi-located nature that conjugates synchronous and asynchronous communications contribute to raising questions about the work, organisational and managerial practices that will favour the co-creation of sustainable work environments. Specifically, Figure 2.2 isolates hybrid work challenges and opportunities for organisations, leadership, and their people, notably when changing from remote work arrangements. Overall, leaders should not only "build and maintain a corporate culture, optimise productivity, work around time zone disparities and develop and invest in security tools to ensure agency and client data is kept confidential" (Poggi 2021, 23), but also support worker well-being and development across places (i.e., office and remote), time (i.e., synchronous and asynchronous communications), and spaces (i.e., physical and virtual) (Gratton 2021; Halford 2005).

With consideration of Gen Z characteristics, values, expectations, and remote work difficulties aforementioned in Section 2.2.1, hybrid work has been purported to present clear advantages over remote work for the youngest working cohort, not least: (1) an improved work-life balance, (2) more frequent in-person collaboration and social interactions, and (3) office spaces that foster interactions, "social cohesion and support work culture" (Pataki-Bittó and Kapusy 2021).

2.3 Theoretical Framework

The context discussed emphasises the necessity to study Gen Z workers' expectations about hybrid work because it will help employers address Gen Z needs, demands, and preferences in a world where hybrid work is rising as the "future of work", blending physical and virtual spaces, overcoming time and place boundaries, and ultimately connecting people to deliver on their jobs. Extant research on Gen Z has focused on students and may be outdated amid evolving work arrangements and practices (Barhate and Dirani 2021; Gabrielova and Buchko 2021; Jayathilake et al. 2021; Pataki-Bittó and Kapusy 2021; Graczyk-Kucharska and Erickson 2020; Schroth 2019), while academic research on hybrid work amid the COVID-19 pandemic is in its early stage and none has especially targeted Gen Z workers.

HYBRID WORK ENVIRONMENT

Opportunities	Challenges
Individual & group level	
• Expanded & diverse workplaces & spaces to support different tasks • Better ergonomics at offices • Increased physical activity at work • More spontaneous collegial support • Work-life balance	• Risk for less autonomy • Decreased individual productivity & lack of motivation to go to the office • Striking the right balance between WFH & the office • Feeling of exclusion when not in the office • Sup-optimal office design • Infection risk concerns • Limited ergonomic support at home
Group level	
• Better conditions for spontaneous meetings, socialising, in depth discussions, creativity, knowledge exchange, & collective learning • Better conditions for building the social fabric, understanding of and connection with each other, & creating a sense of belonging	• Coordination difficulties • Difficult to agree on ways of working • Asynchronised communication can negatively impact on information flow • Invisible groupings between office & remote workers • Lack of office design & meeting culture adjustment for inclusivity in hybrid formats
Leadership level	
• Spontaneity & variation instead of back-to-back meetings; better communication • Capturing what is not seen, made easier than in remote work • Eased onboarding of newcomers • Better work and easier to fulfill work responsibilities • Focus on tasks & performance regardless of workplace* • Office becoming a social hub • Easier international recruitment • Contributing to social & environmental sustainability by combining office & remote work possibilities	• Ambiguities & lack of organisational guidelines & policies on hybrid work • Addressing workers' concerns, providing support for finding new routines & handling frustrations • Clarification of when presence in the office is needed • Implementing rules & guidelines for office usage to ensure safety & productivity • Handling logistics of an uncertain number of workers onsite • Inclusion & equality (ensuring equal opportunities regardless of workplace*) • Creating attractive & cost-efficient solutions for hybrid work

*Workplace refers to the place where work is performed.

Figure 2.2 Hybrid work challenges and opportunities. (Modified from Babapour Chafi et al. 2021.)

To possibly address Gen Z expectations, employers can harness employer branding, whose research remains scarce in the context of hybrid work (Mărginean 2021; Nelke 2021) and which has been most broadly investigated from the attraction perspective (i.e., external employer branding) (Pandita 2021; Pataki-Bittó and Kapusy 2021) rather than the development, engagement, and retention prism that internal branding helps construct.

Figure 2.3 incorporates Gen Z and internal (employer) branding into the theoretical framework of the present chapter. It displays the co-creation of hybrid work between Gen Z white-collar workers and their employers. Remote workers, typically WFH, and those working in the employer's

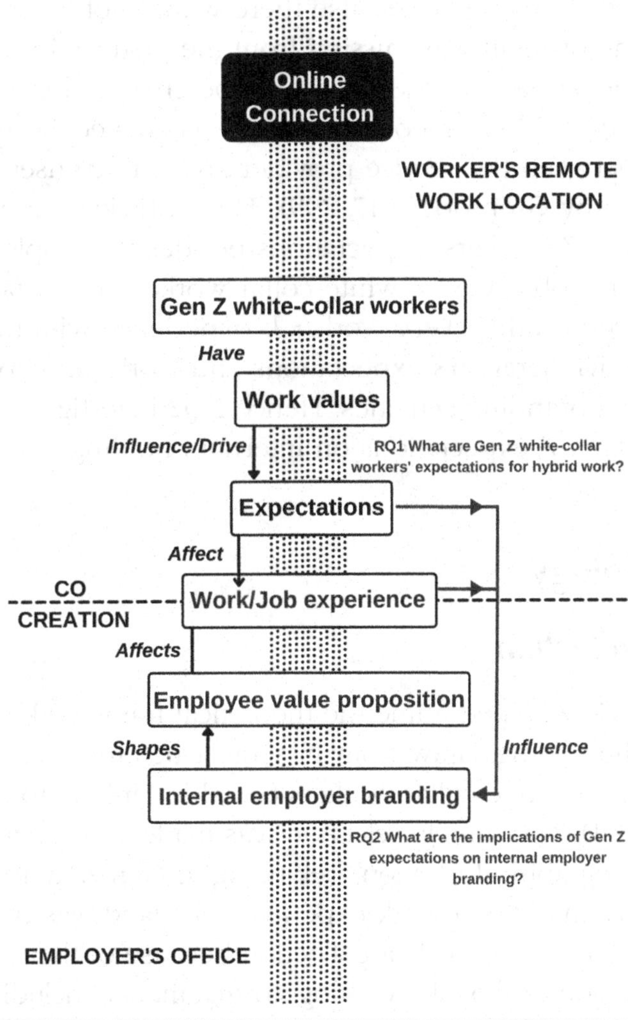

Figure 2.3 Theoretical framework – Co-creating hybrid work for and with Gen Z white-collar workers.

office are all connected in the online realm that bridges the gap between people, time, and places. This online connection is centrally situated because all workers, regardless of their workplace, use technology to communicate and collaborate online. On the one hand, Gen Z white-collar workers have work values that influence/drive their expectations, which in turn affect their work/job experience. On the other hand, internal (employer) branding activities shape the employee value proposition (i.e., the offering intended for the workforce) which affects Gen Z's work/job experience. It is precisely at the level of the work/job experience that occurs the co-creation between Gen Z and their employers: together they mould the work/job experience. Gen Z white-collar workers and their employers are interdependent because there would not be any work/job experience if one of them was missing from the picture. In other words, the value of white-collar work/job can only be created if the work/job is indeed experienced by someone (i.e., value is created "in use", at the meeting point between the service provider and service user, reference value co-creation by Grönroos 2017, 299). The work/job experience in return affects Gen Z workers' expectations besides the employee value proposition. RQ1 probes Gen Z white-collar workers' expectations for hybrid work based on their very work/job experience with their employer. In this research, iGeneration's expectations and work/job experience influence internal branding activities. Hence RQ2 investigates the implications of Gen Z expectations on internal branding.

2.4 Methodology

2.4.1 Data Collection

The literature review helped frame the theoretical framework that guided this research. The findings draw from 15 semi-structured qualitative online interviews of Gen Z white-collar workers based in Finland to collect relevant data (see Table 2.1). The sample selection was made according to four major criteria: geographical location, belonging to Gen Z, holding a white-collar job, and having hybrid work experience. Finland was considered a suitable geographical area for data collection because of the widespread local implementation of flexible working arrangements, including hybrid and remote work, a tendency reinforced by the nationwide governmental recommendation of WFH "when possible" communicated at the end of the

Table 2.1 Gen Z Interviewees

Interviewee Number	*Age*	*Gender*	*Location*	*Workplace*
Z1	25	Female	Tampere	Microsoft
Z2	25	Male	Helsinki	Social Democratic Students of Finland
Z3	25	Male	Kuopio	Pina Oy
Z4	25	Female	Espoo	EP-Logistics
Z5	22	Male	Hämeenlinna	Anonymous
Z6	22	Female	Oulu	DNA
Z7	24	Female	Jyväskylä	UPM
Z8	24	Female	Turku	Yleinen Työttömyyskassa
Z9	25	Female	Tampere	Lyyti Oy
Z10	24	Female	Helsinki	Tampere University
Z11	21	Female	Tampere	A-Lehdet
Z12	25	Male	Vaasa	University of Vaasa
Z13	25	Male	Tampere	Hyperkani
Z14	21	Male	Helsinki	Anonymous
Z15	24	Female	Tampere	Anonymous

year 2021 (Finnish Government-Valtioneuvosto 2021). Priority was given to having interviewees living and working in the same country to benefit from the same macro-environmental (politico-economic, socio-cultural, and environmental) context. Interviewees' age ranged from 21 to 25; nine females and six males, all doing knowledge work, namely white-collar work, at least on a part-time basis and they were all familiar with the hybrid working model. All online interviews were completed via Zoom video calls that were audio-recorded with interviewees' consent and transcribed manually. They spanned three weeks from the end of December 2021 through mid-January 2022. The average interview duration was 44 minutes and 27 seconds. Qualitative interviews resemble conversations guided by the interviewer to capture the meaning conveyed by the interviewee. They were semi-structured to unveil "deep information and understanding" of the sample's viewpoints and experiences (Johnson 2001, 106).

2.4.2 Data Analysis

Interview data were coded inductively employing ATLAS.ti to pinpoint "patterns and relationships between variables and themes" (Given 2008, 121). Categories that emerged from the code were put into perspective with extant literature, notably Gen Z values (see Figure 2.1) and hybrid work opportunities and challenges (see Figure 2.2), to develop findings for RQ1 and RQ2. Figure 2.4 displays the value categories inductively formed and differentiated from Gen Z interview data. Interview data were first grouped into categories that were subsequently associated with an overarching value set encompassing Gen Z expectations. In total, 47 codes were derived from the interviews and eight principal Gen Z values were distinguished, namely:

- " Ethics and Fairness"
- "Leadership"
- "Learning and Development"
- "Lifestyle building and Wellbeing"
- "Purpose and Ownership"
- "Recognition, Rewards, and Benefits"
- "Collaboration and Team spirit"
- "Relationships and Social cohesion"

Ethics and Fairness	ESG; DEI; Intentional design & arrangement of the office; Person-organisation fit; Consistency in actions.
Leadership	Caring supervisor and management; Mentorship & Guidance; Leader training
Recognition, Rewards and Benefits	Appreciation; Office perks; Performance evaluation; Perks & Benefits; Recognition
Learning and Development	Continuous learning & development; Two-sided feedback system
Lifestyle Building and Wellbeing	Commute; Flexibility; Ideal work week; Ideal workday; Lifestyle building; Personal wellbeing; Rare office visits; Rejuvenation; Remote work first; Self-discipline; Stress & Apprehension; Time management; Tracking wellbeing
Purpose and Ownership	Focus & Distractions; Freedom & Ownership; Interest & Drive; Out-of-the-box thinking; Prioritisation; Problem-solving; Purposeful; Responsibility; Self-Awareness; Self-Leadership
Relationships and Social cohesion	Collaboration; Communications; Hybrid meetings; Support & Team Spirit
Team spirit and Collaboration	Community & Networking; Cross-channel communications; Hybrid events; Social connection & Relationships; Trust

Figure 2.4 Gen Z values.

Gen Z expectations relative to these values are presented in the following section to respond to RQ1. Finally, RQ1 findings were instrumental to isolate RQ2 internal branding implications, made prominent in Section 2.5.

2.5 Findings: Gen Z Expectations for Hybrid Work

This section exposes Gen Z's expectations for hybrid work (ref. RQ1) as understood and elaborated on by the researcher. Findings indicate the relative Gen Z proportion who reported experiencing them in-between brackets while direct quotations anonymously refer to interviewees by means of their assigned number preceded by the letter "Z", e.g., "Z1" refers to the first interviewee.

Gen Z's expectations for hybrid work are presented and developed referring to their values identified during interviews (see Figure 2.4). First and foremost, Figure 2.5 hereafter recontextualises these values in the hybrid work landscape, composed of physical workplaces and online workspaces. This demonstrates that Gen Z white-collar worker values and unfolding expectations span across the physical and online realms, thus meaning that both exist throughout the Gen Z work/job experience, irrespective of their presence in the employer's office or remote work location. Gen Z expectations are listed below following their values' alphabetical order, although they were attributed a colour code that will ease the classification of their relative expectations in the following stage of interpretation.

With regard to the ethics and fairness value, five main codes arose (see Figure 2.4) and six expectations were referred to during interviews, that is:

1. Feeling safe to express themselves overall and involved in the shaping of organisational values to develop and sustain "a common sense of purpose" (two-third of Gen Z interviewees).
2. Feeling respected and included in the work community (nine out of 15 interviewees)
3. Working on a diverse team (according to four-fifth of the sample).
4. Benefiting from equal access to resources and opportunities (with reference to 11 out of the 15 Gen Z participants) with a low power distance (for two-third of participants).
5. Consistency in organisational communications, between communications and actions (cited by four-fifth) in addition to an agency on the related consequences (mentioned by eight interviewees).

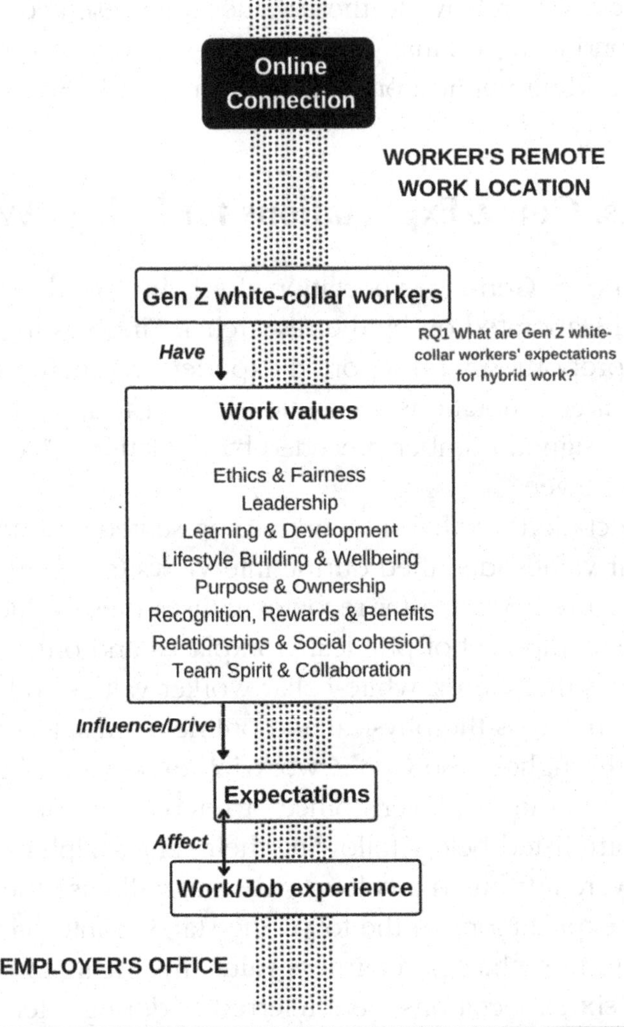

Figure 2.5 Gen Z values in a hybrid work context.

6. Initiatives that contribute to the fulfilment of environmental, social, and governmental (ESG) goals – with at least some degree of importance granted to social or environmental matters (unanimous opinion across all interviewees) – and tracking and communication about the evolution of organisational activities over time to remain aware of the progress made towards more sustainability (i.e., better impact) (raised by eight Gen Z participants) as mentioned: "I think both social and environmental aspects matter. It's very important to track changes and evolve, focusing on what can be achieved with sustainability" (Z1).

7. Importantly, "there should be the same goals and principles in hybrid work as when working in the office" (Z12) for all workers to be equally treated, supported, evaluated, and rewarded when working towards the achievement of their goals.

The leadership value arose from three main codes (see Figure 2.4) that emphasised two trust-based expectations:

1. Having caring management and supervisors, who have a human approach to leading, show understanding of individual situations, and assist people where and when needed (as discussed by four-fifth of the sample) as epitomised: "I think I have a pretty good employer, very flexible, and I think they really care for their employees. They invest in our education and different kinds of training programs. Plus, we have access to some mental health services and different employee activities" (Z6).
2. Having a direct supervisor who guides them, sets the direction and expectations for the team, is inclusive, gives clear instructions, and lets people work autonomously, yet remains present, available, and approachable to provide help; someone with whom they share a trustworthy relationship and who is open to discussion on various thoughts, ideas, and regarding individual development (unanimous to all interviewees) as recognisable from the quote "I expect my team leader to let me know what we are reaching for, what (s)he expects from me because it's easier for me to do my job when I know what I should get done. But maybe the most important thing I see in team leaders is that they're kind of coaches for their teams, helping them navigate different situations" (Z5).

Learning and development is a telltale value that sprang from two codes and stresses two expectations:

1. Receiving proper and sufficient onboarding, including support from a more experienced worker (revealed by two-third of Gen Zs), to get familiar with the organisation, one's role, tasks and tools, one's team (unanimous amongst interviewees) and how to navigate hybrid work as illustrated: "What should be held in Teams online meetings and what in-person meetings? That is at least one topic I would like to learn more about" (Z1), "Remote communications require training" (Z8), "I need to

develop facilitation skills and collaborative skills to work with a variety of experts across physical and online worlds" (Z9).

2. Continuous learning and development opportunities, to feel challenged but not overwhelmed or overworked, be safe enough to make mistakes (stated by eight interviewees), have a two-sided and inclusive feedback system (suggested by two-third) although one-third of interviewees also reported not knowing how to give constructive feedback and, most importantly, two-third expect to be asked how they would like to develop themselves and grow within their organisation.

Perhaps one of the key values is lifestyle building and well-being, the one that generated the most codes (see Figure 2.4) and singled out four expectations:

1. Having the flexibility and freedom to organise themselves, their work (including breaks), and their life the way they want (shared by all interviewees).
2. For all but one interviewee who would prefer fully remote work (Z11), this entails a hybrid work week with two-three days spent WFH and conversely two-three days spent in the office.
3. The establishment of clearer boundaries between work and the rest of their life (concordant between all interviewees) because "there is more to life than work" (reported by half of the sample) such as spending time with their circle, doing activities they enjoy, thus putting forth the importance of sustaining a feeling of control and stability for seven Gen Zs.
4. Receiving their employer's support to continuously sustain their well-being and happiness through free access to special services such as mental health apps (for half of the participants) and ergonomic arrangements (alluded to by eight), being asked how they are doing notably regarding their workload, stress level (specified by half) and what they are missing to feel better; but Gen Z should also personally make sure that they spend time in nature (unanimous call), exercise, take care of themselves to rejuvenate.

Purpose and ownership gathered most codes after lifestyle building and well-being (see Figure 2.4) and referred to three unanimous expectations:

1. Feeling interested in the job and driven to execute well on meaningful tasks that foster creative problem-solving.

2. Individual authority (i.e., responsibilities) and assurance through autonomous/independent decision-making, self-led endeavours, prioritisation (for which a third needs training), as well as holding people accountable for their actions or inaction.
3. Feeling a person-team and person-organisation fit that feeds their sense of meaningfulness, engagement, and commitment.

The recognition, rewards, and benefits value highlighted two core Gen Z expectations, that is:

1. Basic healthcare benefits and appreciation for their work in the form of a "thank you" (common to all interviewees).
2. Equal work recognition between all workers not only through financial, material, or time benefits such as a pay raise or bonuses (praised by seven Gen Zs), sports/cultural offers (appreciated by four-fifth) or paid time off (celebrated by a fifth of the sample), but also through offers to advance in another position that suits them (proclaimed by 11 out of 15).

Relationships and social cohesion also utterly matter to Gen Z interviewees, who all expect:

1. To maintain sufficient, good, and casual/relaxed social connection to their work community, especially during breaks on their workdays.
2. To build trustworthy relationships with their co-workers.

Furthermore, and bound to relationships and social cohesion, team spirit and collaboration is the third most represented value amongst Gen Z participants, who consensually expect:

1. To belong to a strong and supportive team that has an open culture, where they get along well with their teammates, and everyone has each other's back.
2. To communicate with their teammates daily via the most appropriate collaborative online tools provided by their employer, whereby they can have hybrid meetings that enable coordination within and across teams.

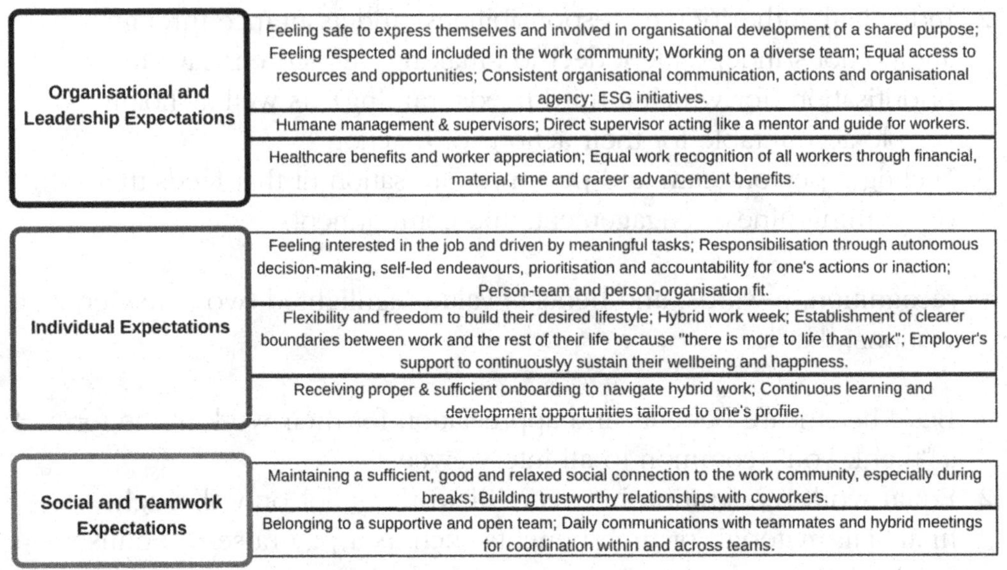

Organisational and Leadership Expectations
Feeling safe to express themselves and involved in organisational development of a shared purpose; Feeling respected and included in the work community; Working on a diverse team; Equal access to resources and opportunities; Consistent organisational communication, actions and organisational agency; ESG initiatives.
Humane management & supervisors; Direct supervisor acting like a mentor and guide for workers.
Healthcare benefits and worker appreciation; Equal work recognition of all workers through financial, material, time and career advancement benefits.
Individual Expectations
Feeling interested in the job and driven by meaningful tasks; Responsibilisation through autonomous decision-making, self-led endeavours, prioritisation and accountability for one's actions or inaction; Person-team and person-organisation fit.
Flexibility and freedom to build their desired lifestyle; Hybrid work week; Establishment of clearer boundaries between work and the rest of their life because "there is more to life than work"; Employer's support to continuouslyy sustain their wellbeing and happiness.
Receiving proper & sufficient onboarding to navigate hybrid work; Continuous learning and development opportunities tailored to one's profile.
Social and Teamwork Expectations
Maintaining a sufficient, good and relaxed social connection to the work community, especially during breaks; Building trustworthy relationships with coworkers.
Belonging to a supportive and open team; Daily communications with teammates and hybrid meetings for coordination within and across teams.

Figure 2.6 Gen Z expectations for hybrid work.

Figure 2.6 encapsulates all the expectations isolated from Gen Z white-collar worker values and reorganises them between three overarching areas:

1. Organisational and leadership expectations comprise (1) ethics and fairness, (2) leadership and recognition, and (3) rewards and benefits values.
2. Individual expectations include (1) purpose and ownership, (2) lifestyle building and well-being, and (3) learning and development values.
3. Social and teamwork expectations cover (1) relationship and social cohesion and (2) team spirit and collaboration values. This taxonomy underpins the turn to internal branding implications that materialise from Gen Z's expectations for hybrid work.

2.6 Internal Branding Implications for Hybrid Work

Drawing internal branding implications for hybrid work becomes possible in light of Gen Z white-collar workers' expectations. As a recall (see Figure 2.3), internal branding aims at establishing and sustaining trust in the context of work/job experience co-creation between workers and their employer. Figure 2.7 presents Gen Z's expectations for hybrid work, here instrumentalised to enhance Gen Z work/job experience by adjusting the

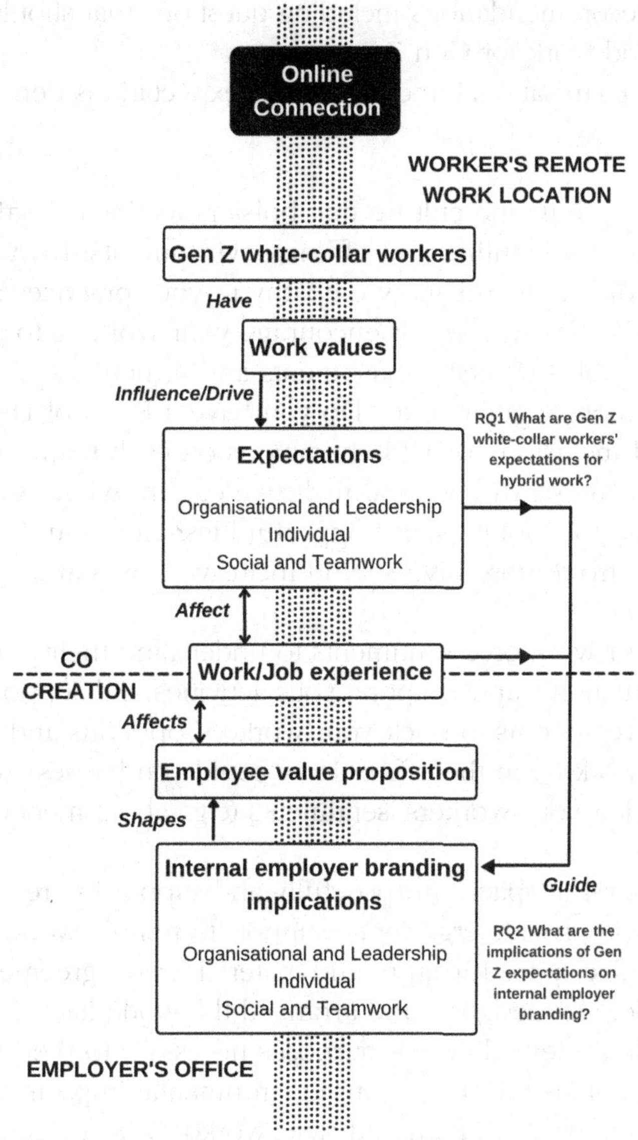

Figure 2.7 From expectations to recommendations for hybrid work.

employee value proposition offered to them, thereby advancing reinforced trust to the extent that employers carefully strategise and operationalise their internal branding activities based on the following implications. These implications for hybrid work stemmed from the researchers' understanding of expectations presented in Figure 2.6 and they conform to the same threefold structure: (1) organisational and leadership implications, (2) individual implications, and (3) social and teamwork implications. All of them come as recommendations including questions that should help orchestrate hybrid work for Gen Z workers.

Specifically, organisational and leadership expectations denote five major implications:

1. Establishing a humane culture that bolsters feelings of safety and acceptance across online and offline environments: Do you have a humane dialogue and show empathy in your practices and behaviours? Do you share or encourage your workers to genuinely share stories of their accomplishments and difficulties faced on a weekly or more regular basis? Do you have a Head of DEI (Diversity, Equity and Inclusion), or DEI champions on each team, who could assist in creating a diverse and inclusive culture where everyone has equal access to resources and opportunities? How much value could you extract from more diverse and inclusive teams and what would be the benefits?
2. Probing your workforce sentiments to understand their purpose and peculiar situations, and adapting your activities and support accordingly: Do you have systems to track your workers' opinions and feelings about WFH and working in the office? How could you harness weekly check-ins to monitor your workers' sentiments (e.g., about mental and physical health)?
3. Devising physical spaces purposefully and supporting remote workers: How can you provide ergonomic support to remote workers? Could you invest in office space furniture and material lease agreements? Have you intentionally designed and arranged the workplace to not only help people achieve different purposes necessary to their work and well-being but also improve your environmental impact? How can you leverage technology to plan and steer workdays and weeks?
4. Elevating mentorship to improve Gen Z workers' experience and development: How can you ensure that every newbie or early-stage worker has a mentor who can help them navigate the work-scape? When should they change mentors?
5. Systematically appreciating and recognising Gen Z workers' efforts, development, output, and peculiarities: How can organisational and team leaders keep track of workers' progress or lack thereof? What could be done for better recognition? Have you asked Gen Z workers how they would like to be recognised and rewarded for their work?

Individual implications revolve around:

1. Empowering and challenging your Gen Z workers to catalyse a virtuous circle: What should on boarding encompass to help them thrive on hybrid work (e.g., time management, focus, project management, and online communications)? What connected tools should you adopt and train workers for to keep everyone up to date in a timely fashion? How can you keep them interested and help them find a sense of accomplishment? How can you help them prioritise, delegate, or refuse work tasks and projects? Do you have weekly check-ins to estimate the progress made? Do you know what their short-, mid-term, and long-term professional ambitions are? How can you turn these into inspiring, "actionable, measurable, actionable, realistic, and time-bound" (Chan et al. 2018) goals they can consistently progress towards and support them in the desired direction?
2. Encouraging and helping Gen Zs build their desired lifestyle through flexibility, coordination, and connection: What requires whose presence in the office and at what frequency? Conversely, what can be performed remotely? How can you implement weekly individual well-being check-ins for all and how to use key data points to assist in the correction of potentially detrimental tendencies? What technological and material tools can you use to foster a greater level of well-being? Do you encourage breaks at least ten-minute long between meetings?

Social and teamwork implications notably entail:

1. Continuously educating, training, and developing Gen Z workers so they can all use the same tools to stay connected and collaborate: What are the key technological channels you want your workforce/team to use, how and why? Does everyone in your ranks/on your team know how to use them and act accordingly? What other channels could you consider improving communications and collaboration across physical and online realms? What subject requires synchronous interaction and what, on the contrary, can be dealt with asynchronously? Do you have official time windows during which you strongly recommend switching off from work, or at least from online communication channels?
2. Mastering hybrid events for targeting, engagement, inclusion, and networking: When are hybrid meetings needed, for whom, and how to ensure all participants can express themselves, be heard equally

and relevant workers can access the meeting minutes/notes/recording? What type of weekly hybrid events would people want to freely join in, irrespective of their work location? How to use which technology to create immersive and inclusive hybrid events? Could you organise hybrid co-working sessions (even across teams) twice or thrice a week, so everyone develops a common sense of work, builds, and activates weak social ties, and shares and receives the help needed?

3. Instilling an open, supportive, and genuine culture that stimulates team spirit: Do you encourage daily interactions, whether it be an engaging post in a group chat, going for lunch altogether, or personally asking others on their team how they are doing? Does everyone on each of your teams know their teammates, what they are responsible for, and who they could refer to for specific skills/tasks/activities? Likewise, does everyone have access to an online list of workers, their affiliation, their current work responsibilities, their availability for giving help as well as for a chat/meeting? Do people know each other well enough to talk about other matters than only professional ones with their teammates? How to make sure they do? Do you ask your workers what common activities they want to do with their teammates to rejuvenate during breaks? Do you allocate a certain budget for team building activities and tools?

Figure 2.8 outlines all these key recommendations for hybrid work based on Gen Z white-collar workers' expectations.

Because hybrid work implies a redefinition of work, an adaptation of existing practices and the adoption and implementation of new ones, it requires a mindset shift that focalises on workers and drives decisions accordingly. Hybrid work is multi-faceted and human-centric, thereby calling for flexibility, openness, authenticity, and coordination to not only co-create the most sustainable work/job experience for each worker individually, but also co-create a shared hybrid work experience for all workers together, ultimately improving organisational morale, culture, commitment, performance, and attractiveness. Technology undeniably enables this co-creative work across time and space boundaries and finding the most suitable tools to help organisations and individuals reach their goals is, therefore, an imperative. Experimentation completely fits this new paradigm of work, where nothing is finite and everything can be morphed according to the localised needs, goals, preferences that arise over time.

HYBRID WORK RECOMMENDATIONS

Organisation and Leadership

- Establish a humane culture that bolsters feelings of safety and acceptance for all across online and offline environments
- Probe your workforce sentiments to understand their purpose and peculiar situations, and adapt your activities and support accordingly
- Devise physical spaces purposefully and support remote workers
- Elevate mentorship to improve Gen Z workers' experience and development
- Systematically appreciate and recognise Gen Z workers' efforts, development, output and peculiarities

Individual

- Empower and challenge Gen Z workers to catalyse a virtuous cycle from onboarding onwards
- Encourage and help Gen Z workers build their desired lifestyle through flexibility, coordination and connection

Social and Teamwork

- Continuously educate, train and develop Gen Z workers so they can all use the same tools to stay connected and collaborate
- Master hybrid events for targeting, engagement, inclusion and networking
- Instill an open, supportive and genuine culture that stimulates team spirit

Figure 2.8 Hybrid work recommendations to improve Gen Z workers' experience.

2.7 Research Limitations and Future Research

This chapter delved into Gen Z white-collar workers' experience of hybrid work and their relative expectations to apprehend the internal branding implications for organisations. It stands amongst the first studies that explore this subject amid the COVID-19 pandemic. The findings converge with previous work on Gen Z student features, not least in the context of job search, thus accentuating their generational profile driven by strong values and related search for humane work environments, ethics and fairness, continuous development, lifestyle building and well-being, purpose and ownership, recognition in addition to a strong community. This suggests Gen Z value constancy despite the pandemic and draws attention to the ways values unfold and how they can be fulfilled in hybrid work, where terms and conditions remain relatively malleable.

Future research could focus on one of the three types of implications that emerged from this work, or one of the recommendation points specifically highlighted in Figure 2.7, to deepen understanding of Gen Z expectations' implications on organisations and their leadership, on Gen Z individuals themselves or on social and teamwork aspects. Perhaps this endeavour could gain in accuracy if enquiring them in tandem, namely considering their relationship and mutual influence, if any. This study investigated Gen Z white-collar workers who hold different positions across various organisations in Finland. Thus, this research could be replicated in another country where hybrid work is also implemented for white-collar workers, for instance, one of the Nordics. Moreover, studies targeted at a determined role across organisations would help gain more thorough and targeted Gen Z insights. Likewise, longitudinal individual organisation case studies would be valuable to further comprehend how hybrid work is co-created, that is, the mechanism(s), between different actors across the organisation over time. A promising project could reproduce this work with older working cohorts. Another would explore the fascinating potential generational divide of expectations and associated internal branding implications of hybrid work. Additionally, future work could examine the adoption, utilisation, and relative impact of augmented reality and virtual reality technologies on hybrid work, white-collar workers, and their organisations. In terms of internal branding, this research holistically drew implications from Gen Z hybrid work experience, which implies that further and narrower research avenues could be directed towards specific internal branding phenomena that contribute to shaping hybrid work/job experience.

Now that hybrid work constitutes a major and growing working model, it deserves strengthened attention and interest from: (1) practitioners, notably employers who want to remain or become a great place to work, with subsequent benefits on talent attraction, (2) researchers who aim to uncover the what, why, how, where, and who of hybrid work, (3) regulators, who not only set the employment policies and laws that dictate what can/cannot be done, but can also launch programmes to incentivise sustainable work behaviours, initiatives, measures, and practices. In particular, creating a dialogue with workers and heeding their insights is paramount to co-creating the future of work.

References

Babapour Chafi, Maral, Annemarie Hultberg, and Nina Bozic Yams. 2021. "Post-Pandemic Office Work: Perceived Challenges and Opportunities for a Sustainable Work Environment." *Sustainability (Basel, Switzerland)*, Vol. 14, no. 1: 294. https://dx.doi.org/10.3390/su14010294.

Barhate, Bhagyashree, and Khalil M. Dirani. 2021. "Career Aspirations of Generation Z: A Systematic Literature Review." *European Journal of Training and Development*, Vol. 46, no. 1/2: 139–157. https://dx.doi.org/10.1108/EJTD-07-2020-0124.

Britton, Diana. 2021. "Hightower Makes Hybrid Work Permanent." *Wealth Management*. https://www.wealthmanagement.com/industry/hightower-makes-hybrid-work-permanent

Chan, Mun Yu, Christian Swann, and James Donnelly. 2018. "Are S.M.A.R.T Goals Really Smart? The Psychological Effects of Goal-Setting in a Learning Task." *Frontiers in Psychology*, Vol. 9. https://dx.doi.org/10.3389/conf.fpsyg.2018.74.00020.

CIGNA U.S. 2018. Loneliness Index National Report, Cigna Intellectual Property, Inc.

Deckert, Andrea. 2021. "Financial Industry Embraces Hybrid Work Model." *Rochester Business Journal (Rochester, N.Y.: 1987)*, Vol. 37, no. 27: 8–9.

Deloitte. 2021. "A call for accountability and action — The Deloitte Global 2021 Millennial and Gen Z Survey." https://www2.deloitte.com/content/dam/Deloitte/global/Documents/2021-deloitte-global-millennial-survey-report.pdf.

Driscoll, Kathleen. 2021. "How to Maintain Stability in a Hybrid Work Environment." *Rochester Business Journal (Rochester, N.Y.: 1987)*, Vol. 37, no. 11: 21–23.

Finnish Government — Valtioneuvosto. 2021. "Information and advice on the coronavirus — Current restrictions." https://valtioneuvosto.fi/en/information-on-coronavirus/current-restrictions.

Gabrielova, Karina, and Aaron A. Buchko. 2021. "Here Comes Generation Z: Millennials as Managers." *Business Horizons*, Vol. 64, no. 4: 489–499. https://dx.doi.org/10.1016/j.bushor.2021.02.013.

Given, Lisa M. 2008. *The SAGE Encyclopedia of Qualitative Research Methods*. https://dx.doi.org/10.4135/9781412963909.

Graczyk-Kucharska, Magdalena, and G. Scott Erickson. 2020. "A Person-Organization Fit Model of Generation Z: Preliminary Studies." *Journal of Entrepreneurship, Management and Innovation*, Vol. 16, no. 4: 149–176. https://dx.doi.org/10.7341/20201645.

Gratton, Lynda. 2021. "Four Principles to Ensure Hybrid Work Is Productive Work." *MIT Sloan Management Review*, Vol. 62, no. 2: 11A–16A.

Grönroos, Christian. 2017. "On Value and Value Creation in Service: A Management Perspective." *Journal of Creating Value*, Vol. 3, no. 2: 125–141. https://doi.org/10.1177/2394964317727196.

Half, Robert. 2015. "Get ready for Generation Z." Robert Half International Inc. https://www.roberthalf.com/blog/the-future-of-work/get-ready-for-generation-z.

Halford, Susan. 2005. "Hybrid Workspace: Re-spatialisations of Work, Organisation and Management." *New Technology, Work, and Employment*, Vol. 20, no. 1: 19–33. https://dx.doi.org/10.1111/j.1468-005X.2005.00141.x.

Hepburn, Michael, Bonnie Burke, Deepa Menon, ja Laura Taylor. 2021. "Dr. Jekyll and Mr. Hybrid: Navigating the Hybrid Work Environment." *BenefitsPRO*.

Hernandez, Joseph A. 2020. "Gen Z Join the Team." *Chemical Engineering Progress*, Vol. 116, no. 5: 20.

Hirsch, Peter Buell. 2021. "Sustaining Corporate Culture in a World of Hybrid Work." *The Journal of Business Strategy*, Vol. 42, no. 5: 358–361. https://dx.doi.org/10.1108/JBS-06-2021-0100.

Jayathilake, Hasaranga Dilshan, Dazmin Daud, Hooi Cheng Eaw, and Nursyamilah Annuar. 2021. "Employee Development and Retention of Generation-Z Employees in the Post-COVID-19 Workplace: A Conceptual Framework." *Benchmarking: An International Journal*, Vol. 28, no. 7: 2343–2364. https://dx.doi.org/10.1108/BIJ-06-2020-0311.

Johnson, John. 2001 "In-Depth Interviewing." In the *Handbook of Interview Research*. Edited by Jaber Gubrium, F. and Holstein, J.A. 103–119. SAGE Publications. https://dx.doi.org/10.4135/9781412973588.

Julka, Samantha. 2021. "Trust Will Be Imperative in Hybrid Work Models." *Indianapolis Business Journal*, Vol. 42, no. 10: 6–6A.

Koop, Avery. 2021. "Chart: How Gen Z Employment Levels Compare in OECD Countries." *World Economic Forum*. https://www.weforum.org/agenda/2021/03/gen-z-unemployment-chart-global-comparisons/.

Lahti, Marko, ja Suvi Nenonen. 2021. "Design Science and Co-Designing of Hybrid Workplaces." *Buildings (Basel)*, Vol. 11, no. 3: 129. https://dx.doi.org/10.3390/buildings11030129.

Lin, Grensing-Pophal. "What Does a Hybrid Work Model Look Like?" Brentwood: Newstex. (2021). https://www.proquest.com/blogs-podcasts-websites/what-does-hybrid-work-model-look-like/docview/2529700751/se-2?accountid=11774.

Mahmoud, Ali B., Leonora Fuxman, Iris Mohr, William D. Reisel, and Nicholas Grigoriou. 2021. "We Aren't Your Reincarnation!" Workplace Motivation across X, Y and Z Generations." *International Journal of Manpower*, Vol. 42, no. 1: 193–209. https://dx.doi.org/10.1108/IJM-09-2019-0448.

Mărginean, Alina Elena. 2021. "Gen Z Perceptions and Expectations upon Entering the Workforce." *European Review of Applied Sociology*, Vol. 14, no. 22: 20–30. https://dx.doi.org/10.1515/eras-2021-0003.

McKinsey. 2021. "It's Time for Leaders to Get Real about Hybrid Work." *McKinsey Quarterly*. https://www.mckinsey.com/business-functions/people-and-organizational-performance/our-insights/its-time-for-leaders-to-get-real-about-hybrid.

Microsoft. 2021. "The Next Great Disruption Is Hybrid Work — Are We Ready?". *WorkLab — The Work Trend Index*. https://www.microsoft.com/en-us/worklab/work-trend-index/hybrid-work.

Nelke, Astrid. 2021. "Impact of the COVID-19 Pandemic on Corporate Employer Branding." *Technium Social Sciences Journal, Technium Science*, Vol. 16, no. 1: 388–393. https://doi.org/10.47577/tssj.v16i1.2436.

Neumayr, Thomas, Hans-Christian Jetter, Mirjam Augstein, Judith Friedl, and Thomas Luger. 2018. "Domino: A Descriptive Framework for Hybrid Collaboration and Coupling Styles in Partially Distributed Teams." *Proceedings of the ACM on Human-Computer Interaction*, Vol. 2, no. CSCW: 1–24. https://dx.doi.org/10.1145/3274397.

Neumayr, Thomas, Banu Saatci, Sean Rintel, Clemens Nylandsted Klokmose, and Mirjam Augstein. 2021. "What Was Hybrid? A Systematic Review of Hybrid Collaboration and Meetings Research." *ACM Transactions on Computer-Human Interaction*. https://doi.org/10.48550/arXiv.2111.06172.

OECD. 2020. "Productivity Gains from Teleworking in the Post COVID-19 Era: How Can Public Policies Make It Happen?" OECD Policy Responses to Coronavirus (COVID-19). https://www.oecd.org/coronavirus/policy-responses/productivity-gains-from-teleworking-in-the-post-covid-19-era-a5d52e99/.

Pandita, Deepika. 2021. "Innovation in Talent Management Practices: Creating an Innovative Employer Branding Strategy to Attract Generation Z." *International Journal of Innovation Science*, ahead-of-print. https://dx.doi.org/10.1108/IJIS-10-2020-0217.

Pataki-Bittó, Fruzsina, and Kata Kapusy. 2021. "Work Environment Transformation in the Post COVID-19 Based on Work Values of the Future Workforce." *Journal of Corporate Real Estate*, Vol. 23, no. 3: 151–169. https://dx.doi.org/10.1108/JCRE-08-2020-0031.

Pearson Higher Education. 2018. "What Do Generation Z and Millennials Expect from Technology in Education?" *Prek-12 Education*. https://www.pearsoned.com/generation-zmillennials-expect-technology-education/.

Pichler, Shaun, Chiranjeev Kohli, and Neil Granitz. 2021. "DITTO for Gen Z: A Framework for Leveraging the Uniqueness of the New Generation." *Business Horizons*, Vol. 64, no. 5: 599–610. https://dx.doi.org/10.1016/j.bushor.2021.02.021.

Poggi, Jeanine. 2021. "Ad World Looks to the Future of Work Post-COVID: 2021 Will Bring New Challenges in Navigating Hybrid Workplaces." *Advertising Age*, Vol. 92, no. 3: 22.

Schroth, Holly. 2019. "Are You Ready for Gen Z in the Workplace?" *California Management Review*, Vol. 61, no. 3: 5–18. https://dx.doi.org/10.1177/0008125619841006.

"Sustainability in the New Remote and Hybrid Worlds of Work: Reduced Workspace Usage Impacts in Indian Technology Firms." *Strategic Direction (Bradford, England)*, Vol. 37, no. 12 (2021): 20–21. https://dx.doi.org/10.1108/SD-11-2021-0135.

Tulgan, Bruce. 2013. "Meet Generation Z: The Second Generation within the Giant 'Millennial' Cohort." Bruce Tulgan and RainmakerThinking, Inc. RainmakerThinking, Inc.125. New Haven. https://grupespsichoterapija.lt/wp-content/uploads/2017/09/Gen-Z-Whitepaper.pdf.

Turner, Anthony. 2015. "Generation Z: Technology and Social Interest." *The Journal of Individual Psychology*, Vol. 71, no. 2: 103–113. https://doi:10.1353/jip.2015.0021.

Weller, Chris. 2017. "A 40-Year Study of Teens Finds Generation Z Is Unlike any Past Generation — Here's What They're All about." *Insider*. https://www.businessinsider.com/generation-z-teens-what-theyre-all-about-2017-9.

Yacine, Leyla. 2021. "Employees-as-Customers: Coupling the Employee Value Proposition and Millennials' Experience in the Construction of the Internal Brand." https://urn.fi/URN:NBN:fi:tuni-202102172132.

WORKFORCE II

Chapter 3

Role of Employees' Social Media Stories in Employer Branding: A Qualitative Study

Mili Dutta

Department of Management, Birla Institute of Technology-Off-Campus Lalpur, Ranchi

Shruti Traymbak

Jagannath International Management School, New Delhi

Meghna Sharma

Amity International Business School, Amity University, Noida

Jimnee Deka

SRF Research Scholar, Amity International Business School, Amity University, Noida

Contents

DOI: 10.4324/9781003267751-5

3.1 Introduction

The perception of a brand is closely associated with the reputation of the organization. Undoubtedly employer branding plays a significant pivotal role in the entire process of hiring as it affects the perception among the applicants and hence a lot on the success of the hiring process. According to Maslow, one of the first things that employers have to ensure is psychological needs like food and sleep (McLeod 2020). Google has free food stores and nap times. These are just a few things out of many that Google does to keep employee satisfaction high. Next, the employer needs to ensure safety needs like security and safety (McLeod 2020). For this reason, a retirement plan is a major incentive for potential employees as seen in government jobs. A good employer does not only provide long-run safety but also a safe work environment. Google's workplace is a big campus calming the mind and ensuring a strong perception of safety. After that, belongingness and love needs need to be addressed (McLeod 2020). Mixed-gender teams boost motivation, and employees perceive them as more alluring than the same-gender teams. A little romance in the workstation does not harm work. Neither does a play zone. Call centres with mixed-gender workstations and child play zones hire better telemarketers. Google has all sorts of recreational activities available that encourage friendship and belongingness. After that, an employer needs to satisfy esteem needs (McLeod 2020).

Dove is a great organization that helps build the self-esteem of its female employees. Finally, self-actualization needs require satisfaction (McLeod 2020). Freedom to work innovatively and creatively is something that most people wish to achieve in their careers. This brings us back to Google as they make millions from the innovation of their workers. Google likes to treat employees as owners of the organization so that they can take the lead (Randstad 2020). After addressing all the needs in Maslow's need hierarchy, the organization is more likely to find that an employee is functioning at an optimum state. They are happy being proficient instead of being unhappy and inefficient.

Speaking about Generation Y, they have grown up with a great amount of familiarity with the internet and are tech-savvy, social media-friendly people. The opinions and views expressed through various social media platforms do have a great amount of effect on their choices. They do refer to the reviews expressed by the previous employees of an organization who have gone through the entire need-based satisfaction process. Visibly, the opinion of existing employees expressed through social media platforms does have a great impact on future applicants. Social media and professional career websites are mostly used by various organizations as a tool to enhance their brand image (Minchington 2014). The Employer Brand International's third global study on employer branding trends, conducted during the period 2009–14, clearly reveals that HR initiatives are responsible for the employer brand. Fifty-eight percent of the activities toward enhancing the brand image are through social media followed by websites and employer marketing activities.

However, it is not enough to just be a fulfilling employer. To alter employer brand perception employees are needed to share their experiences with the rest of the world. For that, employees need to post truthful information about their workplace and the benefits they enjoy on social media. Many social media platforms are free to use and serve the purpose of spreading the good word about employers (Figure 3.1).

Some good examples of social media platforms with global reach include Facebook, Instagram, Twitter, and Snapchat. Organizations that encourage

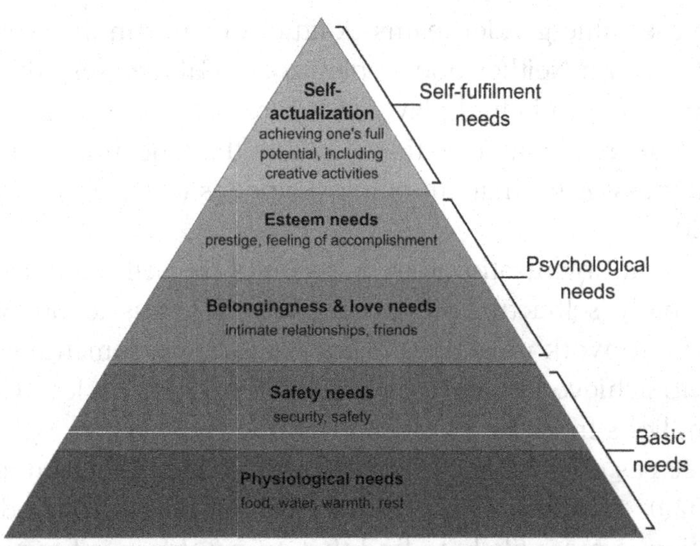

Figure 3.1 Maslow's hierarchy of needs drawn from Wikipedia.

employees to share their stories on social media platforms improve employer brand perception. A great example is Cisco. Their employer brand perception is impeccable. If you search for "best places to work" on Google, Cisco will top the list. This is because Cisco continuously reinvents employer brand perception by incorporating social media into its strategy. Cisco also offers the facility of virtual working. They transform all employees into brand ambassadors giving them creative freedom to post about work. This in turn influences and inspires their network of friends and family to work for the organization. Cisco enjoys outstanding employer brand perception after posts about real-life stories of employees went viral on the internet. The post themselves are simple messages that employees post regularly. It helps in increasing the followers of the organization's social network page and group. The great thing about social media is that the organization can capitalize on the element of instructiveness. Potential employees or just curious people can communicate with the organization through messaging. This feature weeds out people who do not fit in. At the same time, it prepares ideal candidates before recruitment. Thus the organization can choose the best employees out of countless applicants.

The great thing about social media is that it enhances employer brand perception on a global scale. There are no tangible frontiers separating people on social media. The more international recognition the organization gets the better employer brand perception becomes domestically. In this study, it will be observed how big organizations are satisfying the need for employees. Furthermore, the study will see how gratification translates to an increase in employer brand perception with the help of social media usage. The results will show that the incorporation of social media into employer branding strategy will result in the higher successful recruitment of talent. Moreover, the satisfaction of needs is a powerful message to be spread across social media. Giving employees the power to share their experience at work truthfully will lead to an increase in transparency about job facilities. Truth is the greatest influence on employer brand perception.

3.2 Literature Review

Human resource management is considered to be an important department of every organization. It plays a key role in acquiring, developing, maintaining, and utilizing the most crucial resource of an organization – Human Resource. Attracting the best talent in an organization has always been the most challenging job for the Human resource management department. Retaining good employees is also equally important for any organization. Every organization wants to thus create a good image that could help them not only in attracting good employees but also in retaining them. Thus, employer branding is much needed.

3.2.1 Employer Branding

Employer branding refers to an organization's ability to attract the best employees and retain the ones they already have (Backhaus and Tikoo 2004). Being a desirable employer means that more people will want to apply for work. As a result, the employer can pick the cream of the crop during recruitment. With an elite workforce, an organization has a higher chance of being successful in achieving goals. So, to be successful, an organization has to look at what satisfies employees. The more an organization satisfies worker needs the more transparent employer branding perception will be. The needs of a worker are well put in the hierarchy of needs model by Maslow. Organizations that satisfy the need of workers in the order shown in the hierarchy of needs pyramid are most desirable to potential recruits. However, the employer must advertise what it can offer effectively to increase employers' brand perception.

3.2.2 Employer Attractiveness

Employer attractiveness is not how attractive an employer is rather, it is the perception of potential employees and an individual is more likely to apply for a job that he or she sees as attractive (Backhaus and Tikoo 2004). For this, an organization has to advertise itself as an attractive employer. The employer has to advertise the truth otherwise it will not be able to hold on to the employees. Given that the organization is successful at satisfying employer needs, it needs to convey this message to potential recruits. This is done best by employees who are active on social networks. Cisco is a great example of this as they allow their employees to share their stories (Sylvia 2020).

As a result, many such social media posts of Cisco employees have gone viral. Now people from all around the world consider Cisco as their first choice to work. Effective employer attractiveness considers the trends among the generation they are trying to recruit. The message of true satisfaction at work on social media is what attracts employees in today's world.

3.2.3 Needs Hierarchy of Abraham Maslow

The hierarchy of needs is a model that Abraham Maslow came up with. The theory looks at what motivates a human being. Maslow organizes his theory into the famous pyramid diagram that we all have come across. The pyramid has five sections from bottom to top and represents human needs. At the bottom lies physiological needs, next comes safety, followed by love and belongingness and self-esteem needs. Finally, self-actualization lies at the top. Maslow suggests that satisfaction of the human needs from bottom to top will lead to optimum performance (McLeod 2020). Maslow's observation applies to the workplace accurately. When an employer satisfies the needs of workers, they can focus on the work at hand without anything bothering them. Google is a great example of the implementation of Maslow's need hierarchy at the workplace. They put great effort into ensuring that all needs are met starting from physiological to self-actualization. From providing free food to massages, Google does it all. Self-actualization is evident in Google's attitude toward employees. All employees are treated like owners giving them the scope of innovation and leadership (Randstad 2020).

3.2.4 Needs Hierarchy of Abraham Maslow in Organizations

Maslow's hierarchy of needs is a helpful tool that helps us comprehend human behaviour. When it comes to organization, Maslow's need hierarchy shows the best path toward motivation in employees and potential recruits. The five levels in the hierarchy of needs apply to organizations and how they satisfy the needs of their workers. Physiological needs are what need fulfillment; these needs refer to shelter, food, clothing, and other basic needs. To satisfy these needs an organization must give its employees generous wages. A good amount of money buys all the things that an individual may need to survive. So, if an employee provides a wage that is reasonably higher than other organizations the employer branding perception increases. More people are likely to apply for jobs and current employees do not leave their positions easily. Other than a good salary, the employer also has to provide a good work environment as that also falls under physiological needs (Benson and Dundis 2003). Google provides free food, nap times, and even messages to employees and much more. They understand that physiological needs require addressing not just at home but also in the workplace. Furthermore, their salary structure is very attractive. Social media has been a great way for the employees of Google to share their experience at work with the rest of the world. As a result, Google has more job applications than it can handle.

Once physiological needs are taken care of, an employee looks for safety both mentally and physically. For this reason, an organization has to strive toward ensuring a safe environment and job security. A safe working environment increases the productivity of a worker by reducing his anxiety level. Moreover, it is a great incentive to join such an organization. Other than a safe working environment, proper training reduces employees' insecurity. An employee without training is unsure of what he or she is doing. Furthermore, the employee has a higher chance of not accomplishing the task at hand. This reduces job security. On the other hand, an employee with effective training is more likely to work effectively and efficiently. It enhances their work performance and satisfaction. So an organization that invests in the training of employees provides better security (Benson and Dundis 2003). Security is a very attractive aspect of work and increases employer branding perception. The training itself is worth a lot of value to an individual as it will help him or her secure a position in the organization.

Once, physiological and safety needs are taken care of, the need for belongingness requires attention. This is the third stage in the hierarchy of needs and refers to a person's need for love and friendship. Social interaction at work is very important for a human being as work occupies a large portion of the time. There are many other things that an organization can do to increase social interaction in the workplace.

Training is a great place to make friends and relationships that last (Benson and Dundis 2003). Other than training, an organization can offer recreational activities that employees can take part in during breaks. Google has various recreational activities available for employees such as ping pong and billiards. Another great way to increase the sense of belongingness is to promote mixed-gender teamwork. After, the sense of belongingness comes the need for self-esteem in the workplace. Rewards, appraisal, and regular promotions are great ways to promote self-esteem in a workplace (Benson and Dundis 2003). Anything that a worker can gloat about to his or her friends is good advertising for future recruitment too. It increases employer attractiveness.

Finally, the self-actualization need. An organization has to make work pleasurable or addictive for employees. This will increase their confidence and ability to take risks (Benson and Dundis 2003). Google does this well by treating employees like owners. This results in employees taking their work personally and doing the best they can. Furthermore, they have more to brag about on social media furthering employer brand perception among their networks.

3.2.5 Personal Fulfilment in Google

Google is notoriously innovative. This makes the prospect of working at Google very alluring. Google states that personal fulfillment is a very important aspect of employer brand and employer brand perception. The organization treats employees as owners allowing them the scope of self-actualization. Other than that, they are one of the best providers of training. Working at Google means a higher prospect of career progression. The work itself is very interesting and the work environment is wonderful (Randstad 2020). The world gets to know about the amazing work-life at Google from the workers as they post bits and pieces of their life on social media. As a result, Google has become one of the most desirable employers in the world. They get countless job applications regularly. However, Google takes one interview at a tie on a personal level.

3.2.6 Recruitment in Google

Google conveys employer brand to recruit team players who can flourish on a personal level too. Google uses a career website that is much like a social media site. Employees can post their life experiences at Google. This makes it more attractive for potential recruits. The method works and Google gets thousands of applications that they handle one by one (Randstad 2020).

3.2.7 Cisco and Social Media

In 2020, 38.9 thousand people in the USA and 36.6 thousand people around the world began working in Cisco. The number is immense in comparison to organizations across the globe. Cisco has a powerful employer brand that would not be possible without social media. They have a social media policy that allows employees to talk about their life working at Cisco. The employees have become cyber ambassadors of Cisco and thought leaders. They are the gateway to information regarding Cisco. The social media gateway allows potential customers and employees to interact with employees of Cisco (Shaffer 2011).

The employees of Cisco have basic guidelines while posting on Facebook or engaging with people. Firstly, they have to identify themselves as workers of Cisco. They have to abide by policies and uphold the organization's reputation. Next, they cannot make any promises on behalf of Cisco that cannot be kept by the organization. Moreover, they cannot post any confidential information. All posts have to be factual, original, and honest and should add value while building relationships (Shaffer 2011). Cisco won the Rally awards in top career blog and best employer brand video. Furthermore, they took home the prize for being the best in utilizing organic social media and top recruitment marketing campaigns (Sylvia 2020)

Cisco has a social media and employer brand team. They connect candidates and Cisco in innovative ways. By giving their employees consent to generate content on social media, they are making numerous connections. Furthermore, as the content is truthful it is highly influential for potential candidates. They have proven that in this day and age, the best way to increase employer brand perception is through truthful social networking (Sylvia 2020).

By observing big organizations like Google and Cisco it is apparent that the first stage of effective employer branding perception development is satisfying employee needs. To do so, following Maslow's need hierarchy is very useful. The second phase of establishing impeccable employer branding perception is allowing employees to generate truthful content for the global audience on social networks. The result is mass applicants to choose the best from. The creation of employer brand perception in today's world is where the youth is. That place is social media which is also known as the world's eighth continent because of its immense population. The newer continent feeds seven others that in turn provide the creative organizations consumers in millions, many of whom remain truly loyal.

3.3 Motive for Study

The overall motive for the study is to establish the fact that there are many interpretations and scopes for every Maslow's needs. There are a lot of groundbreaking innovative HR strategies to construct from rephrasing the needs in Maslow's need hierarchy. For instance, very few organizations seek to satisfy the need for Gender though it is a primary physiological need that has to be addressed right at the beginning. Other examples of innovation include the employment of the entire family as an employee unit. This addresses the need for "love and belongingness" and "security" perfectly. After all, there is no one better to love you than your loved ones. There is no place safer than home not just for one person but for the entire family. Another possible innovation is micro promotions. Increasing the frequency of promotions with micro advancements in position is very effective in meeting the need for self-esteem. The final innovation I would like to talk about is the self-actualization need. Making employees owners of registered daughter businesses makes them more devoted to the greater cause of the umbrella organization. Why treat employees as owners when you can make them real owners.

Other than providing true satisfaction to employees we also seek to establish the fact that social media participation falls under the need for self-actualization, self-esteem, and love and belongingness; not only that, it is the most efficient tool to increase and improve employer brand perception. This is true because most social networks are free to use. Furthermore, it is very effective given that one single employer has a global friend's network of recruitment age people who he or she grew up with, and has a history of education together. The main objective of the research is to check that employees improve employer branding perception by communicating their job satisfaction truthfully on social media.

3.4 Methodology

3.4.1 Research Design

The first part portrays cases of organizations and how they improve employer branding perception. The present study will identify the need that organizations satisfy and in which category it falls within Maslow's need hierarchy. The second part is to look at the employees of the organizations who are social media users. This study will inspect the prominence of each social presence. Furthermore, we will record views on posts and do a comparative study to see who is influential and who is not. The result supports the hypothesis that social media content generation by employees improves employer branding perception. The specific objective is to show how social media content creation that is truthful and desirable increases employer brand perception.

A qualitative approach is used to answer the question at the core of this chapter. By implementing a case study, the research is going to explore different organizations and see what they do to satisfy their employees. By doing so, one can easily identify what needs in Maslow's need hierarchy are addressed.

The results will be presented in an intra-organization comparison. By doing so, we will see how far the hierarchy of needs is implemented differently by different organizations. The unit of the comparative study is the number of needs satisfied with Maslow's need hierarchy. Next, interviews will be taken with employees of the investigated organizations. A total of ten interviews across four social networking platforms will be used for the study. Each interview is going to be from employees of different organizations included in the investigation. Questions will be asked about whether they post work-related content on social media and how many likes and comments they usually get.

After that, the views on the posts can be counted. The units of analysis are going to be the number of views and comments. Next, a comparison of the posts is going to be made to establish the least to the most effective advertising of an organization by its employee. As a result, the best transmitter of employer brand perception will be found. Finally, it will be seen if there is a relationship between need satisfaction by the employer and employee-generated content reach.

3.4.2 Procedure

Qualitative research with minimum quantitative detail will be used to explore possible innovations that are not applied in today's world of HR. A qualitative explanation is the best way to decode and rephrase the needs of Maslow's need hierarchy. The goal is to achieve the realization that the hierarchy of needs by Abraham Maslow is still unutilized. The Motivational behavioural theory has the potential to revolutionize employer branding. The old model identifies several needs out of which only a few achieve actualization. The employer assumes that money will satisfy all employer needs in most typical organizations. However, that limits the potential of all the needs identified in the hierarchy. The one thing that is more important than money is time. Moreover, work occupies a great deal of time. This period has to be the target and every instant has to be satisfactory for the truly optimum performance of employees. Even this bold and innovative pursuit does not come close to what Maslow's need hierarchy can unlock or uncover.

We are going to partake in two pieces of research and then find a relationship between them. The first research will be for uncovering what needs organizations are satisfying. Researchers are going to collect factual information about what organizations do to improve their employer brand perception. After that, we will see which organization does the most for its employers and how the organizations are benefited from it.

In the second part of the research, we are going to look at the number of views and comments on content generated by employees of the organizations from the first part of the research. We will compare the numbers to see which organization employees managed to reach the highest number of people. The number of views and comments represents message exposure. Message exposure directly correlates to employer branding perception development.

Finally, we make a comparison between the organizations that have been picked for this study. This comparison will show which organization is performing best in terms of employer branding perception.

3.4.3 The Core Questions

The core question of the chapter is a two-step question. In one sentence the problem addressed is: "Does employee-generated content on social media improve employer branding perception if Maslow's need hierarchy is followed to satisfy employer needs?"

3.4.4 Philosophical Foundation

3.4.4.1 Ethics

The basic needs of man merely scratch the true need of man. Abraham Maslow's need hierarchy model came to light in 1943, After a long 77 years, it has not lost its importance as it lets us get closer to finding out what man needs to perform at an optimum level.

A person's life can be divided into two time periods. The two periods are their life at work and life outside work. The objective of Maslow's need hierarchy is to identify all needs in both periods. It identifies that there are five stages of needs that have to be fulfilled in the order of bottom to top. On the bottom, there are physiological needs such as food, rest, and gender. Organizations barely satisfy food and rest needs let alone gender. This shows that employers jump to higher levels of needs before satisfying the basics. This is a great inefficiency in employer branding. Furthermore, it is unethical to keep taboo needs unaddressed.

Social media on the other hand fights the war on unfair workplace needs fulfilment by employers. The reason behind this is that social media facilitates the scope of transparency when it comes to how an employee receives treatment. Employees can truthfully state what he or she is doing in an organization and what the organization is like. This voice not only empowers the employee but also fulfils the need for self-actualization from Maslow's need hierarchy. It also allows the employer to promote the organization as a good employer and improve itself from employee feedback on social media. Freedom of expression on social media by employees is the ethical path forward toward satisfying the needs of man and cultivating opium employees and organizational performance.

3.5 Employee Cases

3.5.1 Overview

For the cases, we are going to look at employees and how they truly feel about their workplace. The testimonials are truthful and may hamper the reputation of the organization. For this reason, the names of the employees and their position titles are going to be hidden from the public eye. One employee refuses to tell us the name of her organization but tells us the general business type. Regardless of confidentiality, employees are truthful about their organization and social media usage with the organization and work lifestyle-related messages.

3.5.2 Approach

We are going to take 10 interviews with employees of different organizations. The interviews come from all around the world. They are going to name the organization they work for. Next, they are going to talk about the needs that their employer satisfies. After that, the interviewed employees are going to mention whether they generate content about their employer and work lifestyle. Finally, they are going to give a rough number of likes and shares they get.

3.6 Results and Analysis

Interview 1 Gender: female Company name: Glasgow royal Usage of social media to improve employer branding: yes Maslow's needs fulfilled: physiological, self-esteem, belongingness, self-esteem, self-actualization The number of likes and comments on average: 50 Interview medium: Omegle	**Interview 6** Gender: female Company name: Bishop Grosseteste University Usage of social media to improve employer branding: yes Maslow's needs were fulfilled: self-actualization, self-esteem, security The number of likes and comments on average: 180 Interview medium: Facebook (this is an interview of a student in training for work; for this reason, an extra question has been added) Ideal employer: Michaela Community school, or Mill Hill County High School
Interview 2 Gender: male Company name: Translate2 Usage of social media to improve employer branding: yes Maslow's needs were fulfilled: security, self-actualization The number of likes and comments on average: 1 Interview medium: LinkedIn	**Interview 7** Gender: female Company name: West 10th eyes Usage of social media to improve employer branding: yes Maslow's needs fulfilled: physiological, safety, and security The number of likes and comments on average: 20 Interview medium: Facebook

(*Continued*)

Interview 3 Gender: female Company name: confidential, a grocery chain store Usage of social media to improve employer branding: yes Maslow's needs fulfilled: physiological The number of likes and comments on average: 15 Interview medium: Facebook	**Interview 8** Gender: male Company name: BRAC Usage of social media to improve employer branding: yes Maslow's needs fulfilled: self-actualization The number of likes and comments on average: 80 Interview medium: Facebook
Interview 4 Gender: female Company name: chitrokarkhana Usage of social media to improve employer branding: yes Maslow's needs fulfilled: physiological, safety, and security. The number of likes and comments on average: 15 Interview medium: Instagram	**Interview 9** Gender: male Company name: C and T Home Care Usage of social media to improve employer branding: no Maslow's needs fulfilled: physiological, self-esteem, and self-actualization. The number of likes and comments on average: 0 Interview medium: Facebook
Interview 5 Gender: male Company name: Rishal group of industry ltd Usage of social media to improve employer branding: no Maslow's needs fulfilled: physiological The number of likes and comments on average: 0 Interview medium: Facebook	**Interview 10** Gender: male Company names: Bangladesh Youth Leadership Center Usage of social media to improve employer branding: yes Maslow's needs fulfilled: physiological, self-esteem, self-actualization, love and belongingness, self-actualization The number of likes and comments on average: 5–10 Interview medium: Facebook

Source: Based on Author's analysis.

The results show organizations are not following the hierarchy of needs from bottom to top. Instead, they skip out on foundational needs. Only one organization managed to hit all levels of the hierarchy pyramid. However, even they didn't satisfy all the physiological needs. This is seen in interview 10.

Employees who post on social media about work get more likes and comments if the employer brand is a globally recognized name as seen in interviews 8 and 6.

University students who are potential recruits have more freedom to post on social networks about the institution and get a greater response. Meaning employer brand perception can be developed from an earlier age than thought in traditional employer branding perception studies.

Employees who do not have permission to post on social media about their employers seem to work at small organizations that have fragile reputations in contrast to concrete employer brands like BRAC.

The number of likes and comments on posts by employees who are female is higher than their male counterparts on average. This suggests females are better employer brand perception promoters on social media.

Organizations that satisfy more needs than other employers still face difficulty implementing employee-generated social media in contrast to organizations that satisfy fewer needs. This shows that smaller organizations are looking into employer-generated content on social media for improving employer brand perception.

3.7 Discussion

3.7.1 Social Media

Females are better mediums of transmitting employer branding perception to the masses on social networks. Brand strength influences social media employer branding perception just as social media employer brand perception influences a brand's strength. These are mutually exclusive.

Finally, the employer brand perception development does not start at the beginning of recruitment or employment. Rather, it starts at university or even earlier. For this reason, university and school students should be encouraged to experiment with brand perception as early as possible using social media. This will increase the following and friend count of students on social networks. This will be handy in moulding employer branding perception later on in his or her career.

3.7.2 Maslow's Need Hierarchy

The organizations that satisfy employer needs most allow employees to post on social media; however, the messages are not effective if the employee and his content do not seem attractive. This is especially true in the case of males. The satisfaction of employee needs is not fulfilled in general according to the hierarchy of needs. As a result, the optimum performance cannot be communicated on social media. This leads to employers banning social media content generation about work.

The question that this chapter seeks to answer is if social media practices by employees improve employer branding perception if they encode a message of satisfaction. The findings show as many as five possible ways of improving employer branding perception.

1. Female employees are more likely to increase employer branding perception than males.
2. Employer branding perception improvement is a skill that can be developed early in life as a student.
3. Employees of strong employer brands are accepted as opinion leaders just as small organizations are made more attractive by employees who are opinion leaders.
4. Maslow's need hierarchy is rarely followed in the order Maslow insists on. Organizations skip basic physiological needs and jump to satisfying needs that are dependent on physiological needs. Following the model will yield better performance for both employees and employers.
5. There is a taboo in giving employees the consent to use social media to improve the attractive, truthful transparency of an organization because of trust issues. Faith in employees by giving them the opportunity of self-actualization through social media freedom increases their reputation as an influence holder in the online community. Even if there is a threat of unsatisfied needs being communicated to a global audience, transparency is the first step toward a concrete foundation. The data shows that even humble institutions that meet few needs of employees allow social media practices as they are confident that transparency is of higher order.

The question at hand is whether an employee's social media communication of truthful messages to a global audience about work satisfaction improves employer branding perception. Our findings suggest that there are many ways in which an employee can improve his or her chances of improving the employer branding perception of the masses. For instance, social media practices from an early age increase the overall reach of the message. It appears that females are natural influence holders and improve employer branding perception better than males. Also, transparency in the form of an employee's freedom to use social networks to describe the employer is respected by the global audience even if all of Maslow's needs are not satisfied. Creative social media usage satisfies the highest order of needs which is self-actualization. So overall, we can say that yes, employer branding perception does improve if employees are treated as cyber ambassadors. However, there is a question of effectiveness that our findings address, like pre-exposure to the concept and gender of the employee. Other than that, the employer has to have faith in transparency for a more effective implementation of the proposed practice.

3.8 Suggestions

3.8.1 Maslow's Need Hierarchy

There are many innovative ways to satisfy the needs of employees, but they are not available yet. The finding highlights that this research suggests that the implementation of innovative concepts such as fluent micropayment and promotion or family unit employment are not yet mainstream. The current situation prevailing throughout the world has made us realize the significance of health first, social belongingness, job security, and upskilling oneself as the priorities of employees. Employees working in organizations or looking for new jobs both are searching for these. During this pandemic, people are more connected through social media and hence before thinking about joining any organization do look for their needs and try to get an assurance through employees' experiences over the various platforms. However, working from home and training are available features in some organizations. Most organizations are now focusing on imparting training through virtual mode, staying connected through remote locations. Still, it is recommended that employers find new and innovative ways to fulfil employee needs from the old hierarchy of needs model in the order Maslow specified.

3.8.2 Social Media

Implementable recommendations based on findings include pre-exposure to the concept of improving employer branding perception through social media content generation. The younger the age, the more time he or she has to gather a mass following. Generation Y has more familiarity with the internet and pandemic compelling people to be more connected through digital platforms. It necessitates that more true and positive stories be shared regarding their experience. social media platforms have developed the immense potential to share information to every corner of the world so connected. Other than that, increasing the number of female employees with the freedom to generate content on social media that is truthful will substantially improve employer branding perception. Next, increasing faith in the transparency of an organization and letting employees freely generate content about work and the workplace will attract potential recruits.

3.9 Limitations and Directions for Further Research

The current situation has posed before us various restrictions because of which more detailed information was not available. First, respondents were limited in number for interviews. The reason being most of the workplaces were closed due to the COVID-19 pandemic, consequently individuals were burdened with managing the personal chores and work engagements. Future research can combine both the qualitative and quantitative research for generalizability of results. Second, likes and comments don't always reflect the employer's brand perception accurately, as there are individuals who are not looking for jobs but who participate in post-interaction. Coincidentally, even their likes and comments influence the perception of potential recruits. Future research shall consider only one organization's employees as a sample for better reflectiveness and shall use other qualitative methods like sentiment analysis. Third, interviews of Google and Cisco employees couldn't be procured to show how the big players in the industry implement employer branding perception, as authors relied on the secondary data available in the public forum. However, authors did manage to get an interview with a BRAC employee.

3.10 Conclusion

The experience of an employee can be a small anecdote and may extend to a strategically fetched story, but should be a true story that can depict the real instances happening in corporates. This will help to give a clear picture of the facts as well as will be able to bring the crowd which is much suited to the needs of the organization. Story telling is thus an efficient way to communicate the needs of an organization thus also helping the candidates yet to join, to understand, and acclimatize with the need-hierarchy existing in the system.

The main question is whether Maslow's need hierarchy guides employers toward providing work conditions that employees can boast about on social media freely and if it increases employer branding perception. The findings suggest that employer branding perception does increase if Maslow's pyramid is kept as a guideline but it is rarely followed to the full extent.

We were surprised to see that our findings suggest that the highest order of needs or self-actualization can be targeted first to get a decent response. But the employer is at risk of negative reviews by employees. Employers should look toward innovation in satisfying employee needs but strictly follow Maslow's need hierarchy so that each layer of the pyramid model is a concrete foundation for the next. Innovations such as family unit employment address a lot of needs all at once. So do fluent micropayment and promotions. These concepts will help organizations leap to the fulfilment of self-actualization that they inefficiently do presently.

Finally, findings show that females are better at improving employer branding perception. However male influence leaders can surpass the average female potential. Furthermore, people should be exposed to social media and the concept of shaping brand perception at an early age. They will have a vast audience by the time they are employees and they will have many tried and tested strategies to implement. For this reason, empowering the female child is the greatest long-run investment that an employer can make for employer branding perception.

References

Backhaus, Kristin, and Surinder Tikoo. 2004. "Conceptualizing and Researching Employer Branding." *Career Development International*, pp. 501–517. 10.1108/13620430410550754.

Benson, Suzanne G., and Stephen P. Dundis. 2003. "Understanding and Motivating Health Care Employees: Integrating Maslow's Hierarchy of Needs, Training and Technology." *Journal of Nursing Management (Wiley)*. 10.1046/j.1365-2834.2003.00409.x.

Chhabra, Neeti Leekha, and Sanjeev Sharma. 2014. "Employer Branding: Strategy for Improving Employer Attractiveness." *International Journal of Organizational Analysis* (Emerald Group Publishing Limited), pp. 48–60. 10.1108/IJOA-09-2011-0513.

Gaddam, Soumya. 2008. "Modeling Employer Branding Communication: The Softer Aspect of HR Marketing Management." *ICFAI Journal of Soft Skills*, Vol. 2, Issue 1, pp. 45–55.

McLeod, S. A. 2020, December 29. Maslow's Hierarchy of Needs. Simply Psychology. http://www.simplypsychology.org/maslow.html.

Minchington, B. (2014). 2014 *Employer Branding Global Trends Study.* https://www.brettminchington.com/single-post/2014/08/05/2014-employer-branding-global-trends-study

Shaffer, L. 2011. Case Study: Cisco Systems, Inc. Open Social Media Policy. https://www.socialmediatoday.com/content/case-study-cisco-systems-inc-open-social-media-policy.

Chapter 4

Challenges to Reinventing Oneself in Contemporary Careers

Karan Sonpar, Federica Pazzaglia, Patrick Gibbons, and Damien McLoughlin
University College Dublin, Ireland

Contents

4.1 Introduction

A popular Harvard Business Case used in several MBA Programmes, Rob Parson at Morgan Stanley (A), presents an intriguing story about a high-performing employee who has received excellent reviews from customers,

DOI: 10.4324/9781003267751-6

met key commercial objectives, and played a key role in enhancing the performance of his unit (Burton 1998). Therefore, it looked like Rob was well positioned to be considered for a promotion to a managerial position. However, a 360-degree feedback revealed mixed views of his performance among his colleagues. On the one hand, they praised his business judgement and acknowledgd his willingness to help them capitalize on opportunities that allow them to cross-sell products. On the other hand, their comments revealed that he tended to work as a lone ranger, and as someone who does not invest time in developing relationships with his peers, something that was problematic in an organization that values a collaborative culture. To make this situation more complicated, Rob was hired to turn around a unit that was not performing too well and, on joining the organization, had been implicitly promised a promotion as Managing Director should he succeed in doing so. Considering his achievements, he was expecting a promotion.

Situations such as the one described in the above vignette are representative of a timeless careers management predicament that arises in organizations. Employees are often unprepared or unable to manage competing expectations and performance requirements within and across different roles and managers are unable to help align their expectations with those of the organizations in ways that support their desired career trajectories. Discussions of the above case during workshops for senior and middle managers typically lead to heated debates among the participants, with some supporting Rob's case for promotion and others arguing against it. Discussions about his manager's actions, however, are generally much less polarizing as everyone agrees that he could have done more to prevent this unpleasant situation, such as by managing expectations and providing timely feedback. However, nearly everyone recalls witnessing similar situations at work.

The above vignette offers a good take-off point for this chapter in which we seek to shed light on the wicked problem of curating one's knowledge and skills through unlearning and adaptation over the course of one's career. A wicked problem has no easy solutions as often solutions can be as disruptive as the problem itself. On the one hand, individuals are told that they are more likely to succeed when they draw on their strengths as doing this allows them to excel in certain areas and maintain a coherent self-narrative. On the other hand, single handedly relying on key strengths independent of one's role and career aspirations can turn out to be counterproductive if one wishes to move up the career ladder. When one discusses this predicament with senior managers, they tend to argue that performance in a certain role, and even when truly remarkable, is an insufficient indicator that one will have a similarly strong performance in the next role. Returning to our opening vignette, they would argue that Rob does not meet the job description for a managerial role, which places a greater emphasis on collaboration, mentoring, and leadership than on individual achievement. Discussions of the same predicament with junior or mid-career managers typically surface perplexity and anxiety about the assumptions they had made about their careers. The takeaway is a simple albeit elusive one, 'what got you here won't get you there' (Goldsmith 2010).

In this chapter, we discuss the ways in which contemporary careers have raised the requirements of adaptation for employees and organizations. We then build on works by career management, leadership, and social psychology researchers to provide insights into how some of the central challenges individuals face as they strive to curate their knowledge and skills as they pursue a desired career trajectory and how these might be overcome.

4.2 The Contemporary Careers Landscape

Researchers across a range of disciplines have drawn attention to the seismic shifts in the ways in which individuals pursue their careers enabled by ongoing trends such as the rapid speed of technological change, global interconnectedness, and the diffusion of new, more fluid, hybrid, and temporary ways of working (De Vos, Akkermans, and Van der Heijden 2019; De Vos, Jacobs, and Verbruggen 2021; Sullivan and Al Ariss 2021). The implications of these trends for individuals are manifold: prior knowledge and technical skills obtained through formal education and subsequent participation in new work and training experiences in the pursuit of career development can become obsolete, established practices and approaches may no longer be relevant, and skills or competencies in one functional area, role, or industry may not no longer be transferable elsewhere. These disruptions, however, also create new opportunities for individuals who are able to reinvent themselves over the course of their careers through adaptation and mobility. This is particularly relevant since the erstwhile dominant concept of a life-time career in one organization (Sarason 1977) has been replaced by careers characterized by a myriad of work experiences spanning roles, organizations, and occupations (De Vos et al. 2019; Sullivan and Baruch 2009).

These changes, often encapsulated in concepts such as boundaryless and protean careers, highlight the fluidity that characterizes contemporary careers. The combination of greater choice and agency by individual workers and ongoing changes in their career landscape make rebooting one's approach to career management more necessary than ever before. However, there is evidence in the academic and practice literature that the pursuit of a desired career trajectory and the adaptation it involves are not so straightforward. We draw on these emerging insights and evidence to present four career traps to adaptation and unlearning and how individuals might overcome them.

4.3 Career Traps that Hinder Unlearning and Adaptation

4.3.1 The Performance vs. Potential Trap

A particularly insidious trap facing contemporary workers is one that is unintendedly created by their organizations as they introduce reward and performance management systems. Academic and practitioner studies highlight how these systems are mainly or solely designed to monitor, assess, and reward the achievements of individuals in their current role as opposed to their potential for future roles. A key implication is that ambitious and achievement motivated individuals are hardwired to engage in behaviours and actions that meet current demands and expectations and these typically lead to actions that are principally targeted at enhancing their own work outcomes such as income and status (Clark, Michel, Zhdanova, Pui, and Baltes 2016; Judge and Kammeyer-Mueller 2012).

However, individuals' pursuit of greater career employability and success often rests on knowledge and skills that are only partially compatible with a dominant focus on attainment and that are not sufficiently emphasized in performance management systems. In their most common form, performance management tools have been argued to breed individualism in organization, a proclivity that is also encouraged by contemporary careers that emphasize employees' agency and choice (Hall 2002; Hogan 2007). This is despite acknowledgements that developing a dependable stream of suitable candidates for higher level positions is an effective strategy for contemporary organizations (Hogan 2007). The assessment of potential instead would require organizations to cultivate in their employees a disposition to understanding and a willingness to advance collective needs and performance that are often in contrast or only partially compatible with their own achievement (Hirschfeld, Jordan, Thomas, and Feild 2008; Sonpar, Walsh, Pazzaglia, Eng, and Dastmalchian 2018), examples being a need for soft skills such as collective leadership, mentorship, and political acumen. Demonstrating advancement potential typically requires a shift in the type of skills that employees cultivate through the work and training experiences they pursue as well as in their perspective, often requiring them to deemphasize their specialist skills to favour more generalist skills and to align their professional development with areas of future industry growth that are often dictated by technological and regulatory change. Advancement potential involves perceived suitability of an individual to take on roles of higher scope and potentially greater complexity and high performance is but the first requirement for them to demonstrate high potential (Fernández-Aráoz, Groysberg, and Nohria 2009). If one were to draw parallels with Kerr's (1975) classic 'On the folly of rewarding "A" while hoping for "B"', one could argue that organizational choices regarding performance management systems can hinder employees' chances for career advancement and lead to discontent and turnover among them and potentially sow the seed for organizational failures in the long term. This could happen, for instance, if high performing and therefore highly mobile individuals were passed over for promotions opportunities or if they were promoted within their organization based on their performance despite their lack of suitability and aptitude for higher level positions.

4.3.2 The Paradox of Success Trap

Andrew Gartner, the legendary CEO of Intel, famously remarked: 'Success breeds complacency. Complacency breeds failure. Only the paranoid survive'. These insights are often echoed in business scholarship, be it in Christensen's (1997) work on disruptive technologies that it is the 'well-run' companies that are disrupted since they get too focused on what they do well and ignore the opportunities at the periphery. Meanwhile, Miller (1992) offers comparable insights in his work 'The Icarus Paradox' in which he argues that companies fail not because they lost their edge but instead because they developed 'too sharp an edge'. While developed with reference to organizations, the social psychology and careers literature note how this predicament applies to individual workers as well and highlight the value of learning over the course of one's career.

There are twin paradoxes around learning. In the first instance, our approach to learning emphasizes the importance of reflecting on experience. As Kolb points out, after we have a concrete experience, we need to reflect on that experience, then conceptualize that experience in exploring causal-like relationships and finally, developing a plan of action to develop a superior response in the future. In discussing this approach with executives, one tends to be regularly confronted by the expressions suggesting that a systematic 'lessons learned' approach is taken after a mistake or an error or a failure, however after success, little explicit learning is done and one just 'moves on'. These insights highlight how individuals are more likely to learn from failure than from success. For instance, success could be serendipitous or driven by factors that lie outside the skills or performance of an individual (e.g., working in high performing teams; corporate culture). So, certainly for practitioners, the view is that we tend to engage in more superficial learning from success and potentially more systematic learning from failure. At the same time, we also know from coaching professionals that the factors that bring success in one's career tend to change over time (Goldsmith 2010). As far back as Katz's reflections on leadership, executives have been thought to rise through the hierarchy, with technical skills becoming less important and interpersonal skills and conceptual abilities becoming much more important.

A second remarkable feature is that while many authors emphasize the importance of creating a learning organization (Garvin, Edmondson, and Gino 2008), this anthropomorphism reminds us that organizations learn through people. While interventions such as experimentation and systematic transfer of learning are identified as key features of such organizations, the quality of the individuals' interpersonal and conceptual skills are critical ingredients. Therefore, the creation of a learning organization needs to focus on the development of individuals in the organization. That development needs to focus not just on the development of new competencies but more importantly on the process of learning itself.

One such example and link is Satya Nadella's emphasis on developing a growth culture at Microsoft, underpinned by the development of individual growth mindsets. Dweck contrasts the fixed and growth mindsets (see Dweck and Leggett 1988). The former assumes that talent is 'innate' and reflects a fixed attribute that is largely uncontrollable by the individual. Individuals with fixed mindsets see failure as personal failing and tend to avoid challenge and change as these may jeopardize performance. Individuals with an incremental or growth mindset see individual attributes as malleable. Individuals with a growth mindset embrace challenges and see failure as an opportunity for learning and improvement.

As a result, while we applaud developments such as initiatives targeted at the creation of learning organizations, the encouragement of experimentation, and the development of psychologically safe environments to enhance learning among employees, we also encourage organizations to seek to develop a growth mindset among individuals who are credible candidates for advancement opportunities particularly to senior roles. One of the principal objectives of career development programmes should be to enhance individuals' learning capabilities through the development of growth mindsets and the concomitant managerial practice of setting learning goals in addition to performance and behavioural goals (Seijts and Latham 2012). Moldoveanu and Narayandas (2022) speak to the importance of developing meta-cognitive skills in training programmes targeted at individuals in or seeking more senior roles. These skills enhance their ability to deal with wicked and ill-structured organizational activities and objectives. We argue that the ability to learn and the attitudes related to the development of growth mindset are examples of such meta-cognitive skills and merit more focused attention from organizations and senior development providers.

4.3.3 Outdated Technical Skills Trap

Wayne Gretzy, the legendary ice hockey player for the Edmonton Oilers, is argued to have remarked: 'A good player plays where the puck is, a great player plays where the puck is going to be'. Individuals are prone to engage in efforts to enhance their current performance and growth potential, whilst ignoring opportunities that are surfacing in the horizon due to emerging developments in their industry. We use the agribusiness sector to make this point, whereby the industry has traditionally adopted a cost-focused approach with several companies only now catching up with the challenges and opportunities of sustainability, digitalization, and food security. The implication for individuals is, they need to align their technical skills with emerging opportunities in the global food chain today. Admittedly, this requires skill revisions and upgradation in future growth areas.

The agribusiness sector, like others, has achieved great success in the last few decades. However, the management skills which enabled this success require updating today owing to threats posed by requirements for greater sustainability and digitization. The challenge in sourcing such skills rests in the competitive strategic tone of the industry and its perception as an unattractive destination for the best talent. This challenging situation has not deterred large firms such as Mahindra, the world's largest producer of tractors, from revising the compensation package of new hires and enhancing their recruitment and development activities to attract and retain talent (Puri 2012). While this approach has many benefits, it also presents several challenges. Among them, are the difficulties in developing individuals who are skilled in domains that require new skills and business approaches. For example, Diageo (Bell, McLoughlin, and Shelman 2014) has experienced several challenges when growing their African business as it needed the support of individuals with key skills and knowledge of the local market. It also faced bigger challenges due to a shortage of employees with key skills in sustainability, climate change, and digital innovation. The Irish food and agribusiness industry has been faster in embracing the importance of developing individuals with core skills in sustainability that were in high demand by its target market (Shelman, McLoughlin, and Pagell 2016). Significant investment has been made in talent development programmes to embed the skills the industry needed but did not possess (Bell and Mary 2010). Doing so has involved an effort to hire potential employees with a desire to drive sustainability in the industry, trained in cutting edge practices, and embedded within the global sustainability community. The seeding of new management potential has contributed to changing the profile of the Irish agribusiness industry.

However, individuals looking at career advancement within or across industries also need to recognize and leverage the disruption occurring in the space of digital differentiation and innovation that emphasize the need for new skills. As an illustration of the magnitude of this phenomenon, the retail grocery market is forecasted to be 25% digital in the next five years. Despite the penetration of digitalization into many industries, the numbers of employees commanding such skills remain low. The reasons for this lie in short-term pressures for profitability that can overshadow organizations' lack of understanding of how to develop employee potential in this regard, as one student shared in a training programme 'you are asking retailers to become digital marketers and they don't know what to do'. Thus, individuals would be well served by pursuing upskilling opportunities in areas where skills are in high demand such as through taking advantage of short-term placements in other organizational units or ad hoc training courses. Occasionally, the shortage of skills can be bypassed by organizations in the short term. For example, the Ocado Smart Platform offers supermarkets the opportunity to buy software which enables them to overcome this management challenge (Alvarez, McLoughlin, and Kindered 2021). For a number of years after this service was launched sales were slow; however over time many large retailers began to use this platform to scale their online presence. While these would appear to be short-term 'fixes' to be used by organizations as they wait for their budding efforts to develop existing employees' skills in these areas to pay off, they nevertheless highlight the risks for individuals wanting to rise to top positions in contemporary organizations.

4.3.4 The Illusion of Limitless Career Opportunities Trap

The growing mantra that one can be anything one wants to be in a world of limitless opportunity inspired by technological advances and globalization has fuelled a belief that individuals' careers have become potentially boundary less (Arthur and Rousseau 1996; Sullivan and Al Ariss 2021; Sullivan and Arthur 2006) and that a perfect career exists somewhere, waiting to be found (Obodaru 2017; Schwartz 2000). Employees have become increasingly likely to pursue desired career paths, often spanning different organizational, industrial, and even national boundaries (Guan, Arthur, Khapova, Hall, and Lord 2019). Notions such as boundaryless and protean careers have placed efforts by individual employees to increase their repertoire of knowledge and skills at the forefront of these efforts to craft an ideal career (De Vos, Jacobs, and Verbruggen 2021; Sullivan and Baruch 2009). However, employees' pursuit of opportunities to acquire new capabilities and become more employable often comes at the cost of letting go of existing capabilities that may not serve them anymore as they seek to further their career prospects.

The importance of unlearning behaviours that are mainly centred on individual contribution and capabilities is particularly critical in technical and highly specialized fields, where selection and performance in entry roles are largely dependent on employees' technical skills and qualifications. It is in these fields that the risks of promoting individuals who are technical experts but possess little understanding or appreciation of motivation, team dynamics, and communication become particularly apparent. This predicament is illustrated in the experience of Google (Garvin 2013), which undertook a lengthy internal study through Project Oxygen in 2006 to find an answer to the deceptively straightforward question, 'do managers matter?' On the one hand, Google was populated with high-performing engineers who valued the organization's innovation mindset and an absence of formal structures and believed that good management was in essence equivalent to being a technical expert and letting team members be. At face value, this setting echoes the insights provided by implicit theories of leadership which posit that employees expect leaders to embody their organization's beliefs and values (Hogg 2001) make them a point of reference for their sense of esteem and identity (Greenberg, Greenberg, and Antonucci 2007). On the other hand, however, the growth of the company had made a flat organizational structure unfeasible, since it led to senior managers being inundated with operational questions and an absence of a 'conveyor belt' of sorts that allowed company decisions to be quickly followed up by execution. Internal surveys of Google employees conducted during Project Oxygen revealed that they were quite unhappy with their managers. Their comments suggested that good management was important, but with the caveat that being a high performer did not imply that one was a right fit for a managerial role. Instead, good management was principally premised on one's mastery of soft skills such as listening, coaching, providing constructive and actionable feedback and empowering others, skills that were not emphasized in the early stages of an employee's career, whereby recruitment efforts are particularly focused on high general cognitive ability and subsequently on the capacity to collaborate as part of a team. This example has a couple of important implications. First, through their recruitment and socialization activities organizations may inadvertently signal those certain behaviours are particularly valued beyond the specific roles being sought. And second, not only are employees increasingly expected to perform in their current roles but also demonstrate they possess the skills and aptitude required for higher level roles, should they seek career advancement. Taken together, the insights of this experiment point to the importance of unlearning, or at a minimum deemphasizing, old skills such as a sole or central focus on individual employees' contributions and learning new skills such as showing concern and attentiveness to team or organization-based outcomes.

4.4 Suggestions for Future Research

This chapter has drawn attention to the need for altering traditional approaches to managing careers considering a changed career landscape. At the broadest level, the four predicaments identified in this paper can serve as fertile pastures for future academic and practice research that elaborates on the challenges and benefits of adopting a more proactive approach to one's career choices. We also present here a few promising areas for future research. First, and in contrast to much focus in the literature that emphasizes the benefits of a focus on expertise and specialization, this chapter has instead presented insights that emphasize the key role played by unlearning and adaptation. Thus, future research could build on these insights to explore a variety of questions including, what triggers and motivates unlearning, how individuals cope with the sense of loss and uncertainty they may experience during this process and tease out when and how unlearning is particularly effective. Additionally, researchers could explore the role of organizational and personality factors that could serve as antecedents of career adaptation.

Second, future research could unpack the role of authenticity in facilitating career success and the relevance of the 'authenticity paradox' proposed by Ibarra (2015) to contemporary careers. While Ibarra (2015) makes an argument for experimentation with provisional selves, role-modeling, and curiosity as enablers of career success, it would be instructive to understand how and why some people are more able than others to embrace and deal with a lack of authenticity as they pursue their desired careers. It would also be interesting to explore how various social groups (e.g., peers, superiors) respond to the personal transformations undertaken by individual employees and elaborate theory on the various facets of authenticity.

Third, although the notion that 'what got you here, won't get you there' is central to this chapter, its relevance to contemporary careers needs to be explored in more detail across roles and occupational groups. There is merit in exploring whether personal success in one's career is beneficial to the organization given that individual motivations to seek advancement in their careers might not be directly relevant to the goals and needs of the collective. Furthermore, there is merit in exploring the role and influence of bystanders, comparative referent groups, and how relational interactions influence the willingness and type of behaviours used by individuals in the pursuit of their desired careers.

Finally, there are also opportunities to understand how industry factors (e.g., speed of technological change) and organizational factors (e.g., slack) influence the extent to which organizations seek and/or support the change in behaviours discussed in this paper. It would also be important to understand how organizational culture, executive preferences, and organizational history influence the effectiveness of these behaviours. Finally, there are opportunities for researchers to more fully flesh out both the behavioural and technical transformations undertaken by individuals in this era of boundaryless careers. This would involve developing and validating psychometric assessment and selection tools that reward individuals with higher tolerance for ambiguity, potential for learning, and resilience.

4.5 Conclusion

Contemporary careers promise potentially unlimited opportunities but are fraught by a host of new challenges for individuals as they pursue their desired career trajectories. We believe it is important and worthwhile to study individuals' experiences in the context of contemporary careers but note that overcoming the traps we discussed in this chapter is necessary for them to unlock the opportunities for career success they seek. Careers and organizational studies stand to benefit from carefully considering how individuals trade off the benefits and challenges of role and career transitions as they navigate the contemporary career landscape.

References

Alvarez, Jose, Damien McLoughlin, and Natalie Kindred. 2021. *Ocado Group: Ready for the Future*. Harvard Business School, Case No. 521-061.

Arthur, Michael B., and Denise M. Rousseau. 1996. "A career lexicon for the 21st century." *Academy of Management Perspectives* 10 (4): 28–39.

Bell, David, and Mary Shelman. 2010. *Pathways for Growth — Building Ireland's Largest Indigenous Industry*. Retrieved from http://www.bordbia.ie/industry/manufacturers/insight/publications/Corporate Publications/Pages/PathwaysforGrowth.aspx.

Bell, David, Damien McLoughlin, and Mary Shelman. 2014. *Diageo: Innovating for Africa*. Harvard Business School, Case no. 514-054.

Burton, Diane. 1998. *Rob Parson at Morgan Stanley (A)*. Harvard Business School, Case no. 498054.

Christensen, Clayton. 1997. *The Innovator's Dilemma*. Harvard Business School Press.

Clark, Malissa A., Jesse S. Michel, Ludmila Zhdanova, Shuang Y. Pui, and Boris B. Baltes. 2016. "All work and no play? A meta-analytic examination of the correlates and outcomes of workaholism." *Journal of Management* 42 (7): 1836–1873.

De Vos, Ans, Jos Akkermans, and B. I. J. M. Van Der Heijden. 2019. *From Occupational Choice to Career Crafting.* The Routledge Companion to Career Studies: 128–142.

De Vos, Ans, Sofie Jacobs, and Marijke Verbruggen. 2021. "Career transitions and employability." *Journal of Vocational Behavior* 126: 103475.

Dweck, Carol S., and Ellen L. Leggett. 1988. "A social-cognitive approach to motivation and personality." *Psychological Review* 95 (2): 256.

Fernández-Aráoz, Claudio, Boris Groysberg, and Nitin Nohria. 2009. "The definitive guide to recruiting in good times and bad." *Harvard Business Review* 87 (5): 74–84.

Garvin, David A. 2013. "How Google sold its engineers on management." *Harvard Business Review* 91 (12): 74–82.

Garvin, David A., Amy C. Edmondson, and Francesca Gino. 2008. "Is yours a learning organization?" *Harvard Business Review* 86 (3): 109.

Goldsmith, Marshall. 2010. *What Got You Here Won't Get You There: How Successful People Become Even More Successful.* Profile books.

Greenberg, Penelope Sue, Ralph H. Greenberg, and Yvonne Lederer Antonucci. 2007. "Creating and sustaining trust in virtual teams." *Business Horizons* 50 (4): 325–333.

Guan, Yanjun, Michael B. Arthur, Svetlana N. Khapova, Rosalie J. Hall, and Robert G. Lord. 2019. "Career boundarylessness and career success: A review, integration and guide to future research." *Journal of Vocational Behavior* 110: 390–402.

Hall, Douglas T. 2002. *Careers In and Out of Organizations.* Sage Publications.

Hirschfeld, Robert R., Mark H. Jordan, Christopher H. Thomas, and Hubert S. Feild. 2008. "Observed leadership potential of personnel in a team setting: Big five traits and proximal factors as predictors." *International Journal of Selection and Assessment* 16 (4): 385–402.

Hogan, Robert. 2007. *Personality and the Fate or Organizations.* Psychology Press.

Hogg, Michael A. 2001. "A social identity theory of leadership." *Personality and Social Psychology Review* 5 (3): 184–200.

Ibarra, Herminia. 2015. "The authenticity paradox." *Harvard Business Review* 93: 52–59.

Judge, Timothy A., and John D. Kammeyer-Mueller. 2012. "On the value of aiming high: The causes and consequences of ambition." *Journal of Applied Psychology* 97 (4): 758.

Kerr, Steven. 1975. "On the folly of rewarding A, while hoping for B." *Academy of Management Journal* 18 (4): 769–783.

Miller, Danny. 1992. "The Icarus paradox: How exceptional companies bring about their own downfall." *Business Horizons* 35 (1): 24–35.

Moldoveanu, Mihnea C., and Das Narayandas. 2022. *The Future of Executive Development.* Stanford University Press.

Obodaru, Otilia. 2017. "Forgone, but not forgotten: Toward a theory of forgone professional identities." *Academy of Management Journal* 60 (2): 523–553.

Puri, Vikram. 2012. "Agribusiness: A great career opportunity for talented people." *International Food and Agribusiness Management Review* 15 (A),: 27–30.

Sarason, Seymour B. 1977. *Work, Aging, and Social Change: Professionals and the One Life-One Career Imperative.* Free Press.

Schwartz, Barry. 2000. "Self-determination: The tyranny of freedom." *American Psychologist* 55 (1): 79.

Seijts, Gerard H., and Gary P. Latham. 2012. "Knowing when to set learning versus performance goals." *Organizational Dynamics* 1 (41): 1–6.

Shelman, Mary, Damien McLoughlin, and Mark Pagell. 2016. "Origin Green: When your brand is your supply chain." *Organizing Supply Chain Processes for Sustainable Innovation in the Agri-Food Industry*: 205–233. http://dx.doi.org/10.1108/S2045-060520160000005017.

Sonpar, Karan, Ian Walsh, Federica Pazzaglia, Miranda Eng, and Ali Dastmalchian. 2018. "Picking the measuring stick: The role of leaders in social comparisons." *Journal of Management Studies* 55 (4): 677–702.

Sullivan, Sherry E., and Akram Al Ariss. 2021. "Making sense of different perspectives on career transitions: A review and agenda for future research." *Human Resource Management Review* 31 (1): 100727.

Sullivan, Sherry E., and Michael B. Arthur. 2006. "The evolution of the boundaryless career concept: Examining physical and psychological mobility." *Journal of Vocational Behavior* 69 (1): 19–29.

Sullivan, Sherry E., and Yehuda Baruch. 2009. "Advances in career theory and research: A critical review and agenda for future exploration." *Journal of Management* 35 (6): 1542–1571.

Chapter 5

The Genius of Craft Brewing: Case of Third Generation Brewing Entrepreneur of Pune

Gautam Bapat
Dr Vishwanath Karad MIT World Peace University

Nimit Gupta
The NorthCap University

Anuj Kumar
Apeejay School of Management

Contents

DOI: 10.4324/9781003267751-7

5.1 Introduction

India is a country having one of the largest agricultural lands probably second in global rank. With 20 agroclimatic zones, India has all 15 types of climatic conditions in the country. Among the 60 known types of soils in the world, India has 46 of them. India comes second in rank when it comes to the production of vegetables and different types of fruits in the world. India is a country that is known for its sizable production of fruits like bananas and mango. In the last decade of the 19th century, food grain output reached a new height. In 2020–21, the Indian government aims to produce 298 MT of food grain ("India at a Glance | FAO in India | Food and Agriculture Organization of the United Nations" n.d.). Figure 5.1 shows the estimated population of India.

Increased pressure on the country's water resources will force policy rethinking and realignment. Desertification and land degradation are also threatening agriculture in the country. The infographic below depicts the present state of Indian crops.

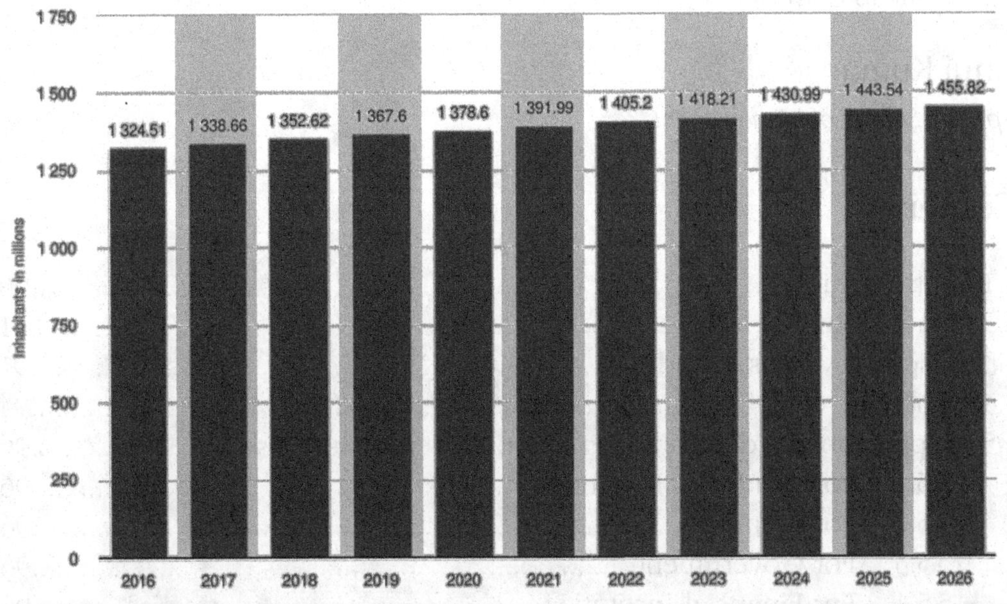

Figure 5.1 Estimated population of India. (Available from https://www.statista.com/statistics/263766/total-population-of-india/.)

5.2 Background

Over the last 50 years, individual Indians' eating habits have changed substantially. The Food and Agriculture Organization (FAO) of the United Nations shows that the country is consuming more all in all. A study showed the consumption trends of nations of the globe for 50 years (1961–2011) ("50 Years of Food in India: Changing Eating Habits of a Rapidly Changing Nation (of Foodies)!" n.d.).

The consumption of animal protein in India has also increased, from 9 grams to 12 grams per day. Around the year 2000, India had a decrease in the number of people classified as undernourished. The previous figure was 210 million, but it has now been reduced to 177 million. Figure 5.2 shows the Indian Agricultural Produce.

It's also been discovered that Indians are consuming more calories these days than they were half a century ago. Although it is assumed that most Indians are vegetarians, it should be emphasized that non-vegetarian diets have also grown in popularity. However, it's worth noting that Indians have unexpectedly upped their fat and sugar consumption.

By 2023, it was predicted that India will see an increase in total alcohol consumption. The total volume is estimated to exceed 7.5 billion litres. It was estimated to be 5.4 billion litres around 2015. Multiple causes, including increased disposable income and a growing urban population, can be blamed for the continuous growth in consumption. Refer to Figure 5.3 which depicts the change in the average daily food consumption of Indians in 1961 and 2011 of these beverages. The Indian alcohol market was divided into two categories: Indian manufactured Indian liquor (IMIL) and Indian made foreign liquor (IMFL). This was on top of the beer, wine, and other imported alcoholic beverages. Country liquor had the largest market share, while spirits accounted for the bulk of consumption.

5.3 Microbreweries in India

Given the environment and customer tastes, India's beer business has been rather unique. The development of microbrewery businesses in India is due to a fast shift in what consumers are searching for and their discretionary cash. People may now drink beer like wine, focusing on its taste rather than chugging it, thanks to improvements in trends in the brewery or craft beer

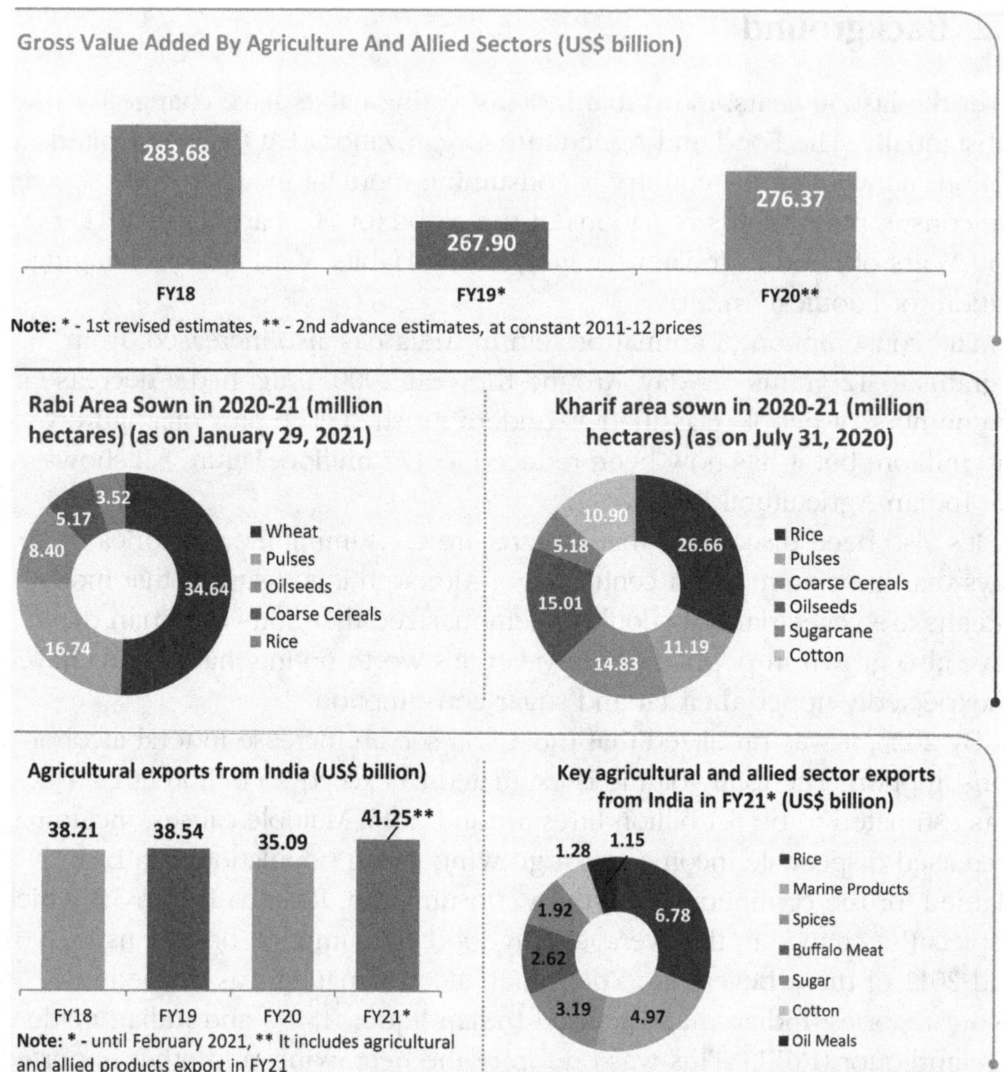

Figure 5.2 The Indian Agricultural Produce. (Available from https://www.ibef.org/uploads/industry/Infrographics/large/Agriculture-and-Allied-Industries-Infographic-August-2021.pdf.)

sector ("The Rise of the Microbrewery Business In India – The Restaurant Times" n.d.).

India is a place where craft beer culture is growing at an incredible rate, and it is unique to the country. The culture has been heavily impacted by European and American models and brands such as Bira and B9, these drinks have grown in popularity in India over time. There is no doubt that

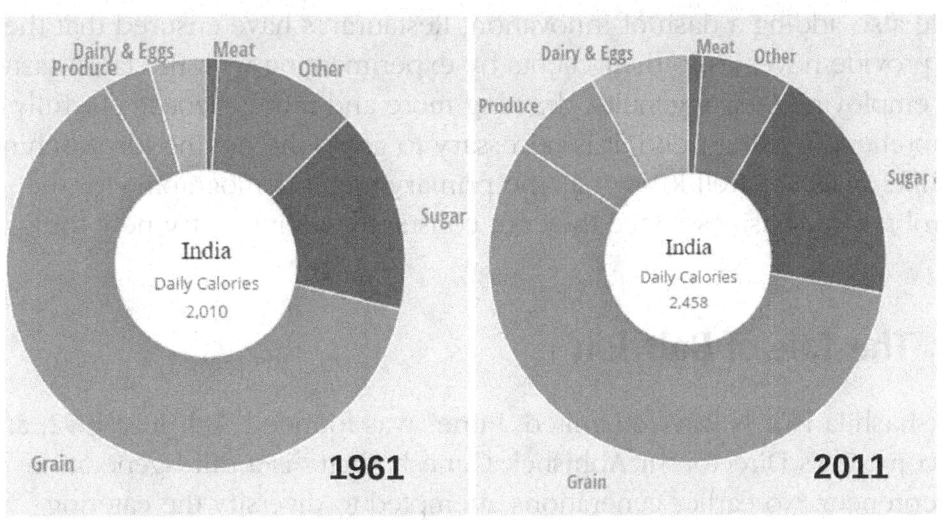

Figure 5.3 Average daily food consumption of Indians in 1961 and 2011. (Available from https://www.thebetterindia.com/98604/india-eating-habits-food-50-years-culture/.)

India's craft beer sector is in its infancy, yet sales of craft beer are claimed to be expanding at a pace of 20% per year across the country.

There is a major "why" behind every movement in the food and beverage sector trend. Aside from the superior flavour and the "buzz," there are a few more factors that have contributed to this industry's unprecedented rise. There are over 2,000 distinct types of beer that may be made. Indians are increasingly looking for more variety and higher quality, which only microbreweries can provide. And, because knowledge leads to experience, Indians, particularly in the metros, now have a greater grasp of beer than they had previously. Following are some of the reasons:

Consumers' proclivity for going out for a beer, and even more so for discovering new places in town, has given the microbrewery industry a boost. The brewery culture is now being incorporated into the menus of future cafés, with fresh craft beer flavours. These new locations have become significant city attractions, resulting in even greater growth for the craft beer sector.

People are constantly late in the big rat race that humans are in, with even less time to go out. To deal with the problem, microbreweries have introduced canned craft beers, which increase mobility. These canned new beer flavours are set to be the next big thing in the microbrewery industry. The microbrewery business is currently staying true to history while also adding a dash of innovation. Restaurants have ensured that they can provide novelties to their clients by experimenting with the latest tastes and employing their ingenuity, drawing more and more clientele. To fully comprehend a rising trend, it is necessary to grasp the driving force behind it. Millennials are well known as the primary target demographic for the microbrewery business since they are constantly willing to try new things.

5.4 The Tale of Babylon

"Takshashila Hotels Private Limited, Pune" was founded 30th July 1992, and the company's Director Mr Abhishek Ganesh Shetty is a third-generation entrepreneur. No earlier generations attempted to diversify the catering industry's verticals. Earlier generations of the family owned and operated a successful restaurant in Pune, Maharashtra. The restaurant franchise had many locations. Mr Abhishek, a third-generation entrepreneur, had the idea of extending the current firm by adding additional services. Mr Abhishek decided to visit Bangalore, where the majority of India's microbrewery sector is situated, to learn more about the numerous options available in the industry. Mr Abhishek was looking for options to open a contemporary style bistro and was intending to buy a new location in the city. After conducting an extensive study online, Mr Abhishek decided to pay a visit to the microbrewery equipment manufacturers to have a better understanding of the industry.

When it came to integrating this new sector into his old firm, Mr Abhishek adopted a methodical approach. He started by looking into challenges experienced by microbrewery entrepreneurs in India on the internet. Then he constructed a systematic questionnaire and distributed it to microbrewery equipment manufacturers as well as microbrewery operators in various areas around the country. After reviewing the responses to this questionnaire, Mr Abhishek devised his list of questions to discuss with the microbrewery equipment manufacturer before making a purchase choice. Because he had strong groundwork done, the equipment makers were now taking Mr Abhishek seriously.

Mr Abhishek then traveled to several areas in Bangalore, Delhi, and Mumbai to observe and research the microbrewery industry. This provided a wealth of information and technical expertise in the operation of the microbrewery. This comprehensive analysis aided owners in determining the total necessary investment, probable profit margins, the expected project timeline, required licences, and raw material procurement, among other things. Licensing, according to Mr Abhishek is the most difficult aspect of the microbrewery operation in Maharashtra. The procedure is exceedingly bureaucratic, time-consuming, and inefficient. Mr Abhishek decided to open a microbrewery after conducting considerable study and doing homework for roughly six to eight months.

Mr Abhishek states unequivocally that the availability of the launchpad was the sole reason for entering the microbrewery industry. Kalinga, one of Takshashila Hotels Private Limited's current restaurants, was one of the top five beer sellers in Pune.

In their restaurant, Kingfisher Draught Beer was one of the best-selling items. The primary impetus for this new company was that, although having top-of-the-chart beer sales, it had a little profit margin. In addition, Kingfisher draught offered just one flavour to all customers. Mr Abhishek's study revealed that customer tastes in Pune are broad and that consumers want items with unique flavours, so he chose to develop flavours as well. The presence of the food industry was the bottom line of this new endeavour, according to Mr Abhishek.

It's worth noting that Mr Abhishek claims that the only reason he was able to succeed in the microbrewery company was that he carried a point of sale with him. One of the most difficult tasks is to set up a microbrewery and then find buyers for the microbrewery's products. The existence of Kalinga, which was already catching a considerable number of beer drinkers, was the most valuable asset. Otherwise, a new company will have to spend money on marketing, sales, transportation, and storage, among other things.

Mr Abhishek stated that most microbreweries in and around Pune city provide a wide variety of beer flavours. He chose to provide six flavours with his microbrewery. In Pune, there are a few microbreweries, such as Effingut, which has 10–15 flavours, and Independence Brewing Company, which has a similar amount. Even though these flavours are well known over the world, Mr Abhishek points out that each microbrewery has its unique formula for creating them.

Mr Abhishek stated unequivocally that taste innovation is one of the primary success reasons for microbreweries. In his microbrewery, he invented a novel beer flavour called Indrayani, which is brewed from locally available rice varieties and is his microbrewery's most popular and successful product.

Mr Abhishek proudly informed us that most of the equipment required for a microbrewery is available in India. But most of the raw materials, like Brewer's yeast, correctly prepared cereal grain (typically barley, corn, or rice), hops, and clean water are all required for making beer. Each component has the potential to influence the beer's flavour, colour, carbonation, alcohol concentration, and other minor modifications. To ensure the best quality, grains are kept and treated with care, most of which are generally imported.

Large amounts of pure water are critical not just as a component, but also for keeping the brewing equipment clean. In beer, water with high levels of lime or iron can disrupt the fermentation process and cause the finished product to discolour. Some brewer's yeast varieties are highly guarded secrets.

5.5 Opportunities due to Rising Microbreweries Business

5.5.1 For Academic Institutions

Academic Institutions can introduce Specialized Courses on Beer Production. Hotel Management Institutions can introduce Beer Crafting as a special course.

5.5.2 For Farmers

Plantation of raw materials is required for microbreweries globally.

5.5.3 For Government

Microbreweries business can prove to be a great revenue resource. The government could consider bringing in ease of doing business for the entrepreneurs.

Regulations towards microbreweries vary widely from state to state. The government could consider bringing in uniformity in the process.

5.5.4 For Financial Institutions

Microbreweries business is one of the fastest-growing businesses globally. The availability of better financing products can benefit financial institutions as well as entrepreneurs.

5.6 Conclusion

In the microbrewery industry, workplace management becomes a key issue. Microbreweries and their output are completely reliant on trained labour. A microbrewery's staff is directed by three persons. The master brewer is one, the production supervisor is another, and the third person might be his apprentice. With just three employees in charge of running the microbrewery, it's vital to look after them and maintain a great balance between production and management.

Abhishek, the owner of the microbrewery, must carefully prepare the goal-setting assessment process and how these employees should be compensated for their efforts. Knowledge management is an important aspect of work and workspace management at any microbrewery because the microbrewery is responsible for producing beer within a certain set of criteria, including location, raw material, and sales.

Mr Abhishek came to a few findings after conducting an extensive investigation. It was obvious to him that a microbrewery may be lucrative if distribution networks are in place. A person who wants to start a typical microbrewery will have to focus on a lot of things in addition to producing high-quality craft beer. Craft beer's quality is scrutinized by a taste-obsessed and exceedingly discerning youthful generation. If your craft beer does not have a point of sale, a lot of time and effort is put into marketing communications and supply chain operations. Craft beer marketing and logistics are both incredibly costly endeavours. The most notable difference between craft and regular beers is that craft beers are typically preservative-free, which reduces their shelf life significantly.

References

"50 Years of Food in India: Changing Eating Habits of a Rapidly Changing Nation (of Foodies)!" n.d. Accessed October 7, 2021. https://www.thebetterindia.com/98604/india-eating-habits-food-50-years-culture/.

"India at a Glance | FAO in India | Food and Agriculture Organization of the United Nations." n.d. Accessed October 7, 2021. http://www.fao.org/india/fao-in-india/india-at-a-glance/en/.

"The Rise of the Microbrewery Business in India – The Restaurant Times." n.d. Accessed October 7, 2021. https://www.posist.com/restaurant-times/features/microbrewery- business-in-india.html.

WORKPLACE

Chapter 6

HR Transformations in Turbulent Times: A Progressive Paradigm Shift

Shilpa Wadhwa

Business Institute, Greater Noida

Parul Wadhwa

EXL, London

Contents

DOI: 10.4324/9781003267751-9

6.1 Introduction

During the last two years of the covid pandemic, the world has witnessed several unseen and unpredicted circumstances of lockdown, layoffs, medical emergencies, and financial crisis which created havoc in every aspect of our daily life. Many nations, organizations, groups, and individuals have realized that the way they worked in the past is not the way of working now and this triggered them to change – either by force or by choice. Experts stated that there will be a drastic change in the work, workforce, and work process in the long run. To accept and manage this covid-forced change, the employers and employees across the globe have strongly joined hands and minds together to transform the work process and policies with the help of the right technology and people support.

Human resource (HR) is just like the internal nervous system of the organization as it takes care of the employees spread across all the levels, departments, locations, nations, etc. The role of HR starts much before an employee joins and continues to be connected with the employee even after he exits the job. Therefore, the time during the pandemic has accelerated the prioritization and transformation of key HR processes and policies for the benefit of its employees and other stakeholders, thereby benefiting the organization at large. Needless to mention, most organizations have realized that HR are the lighthouse during the dark agile times that show direction for innovative ways of working in collaboration and creating an opportunity for a win-win situation to sustain and thrive in the future.

6.1.1 Turbulent Times

The first wave of COVID-19 engulfed the world, where everyone was in a panic and state of shock as to how to face this virus that multiplied and mutated into many of its advanced variants. The concern regarding the pandemic was at a higher level for government authorities, medical experts, and of course the business houses which were struggling to find ways to save the lives of people by providing financial aid, medical facilities, and providing necessities of life. In the struggle to meet these challenges and fight the virus wherein the immediate and safest solution was a nationwide lockdown, the life of everyone came to standstill with no clue of the future; the only concern was safeguarding themselves and their families.

The very question of survival was pondered over in the minds of people. As the saying goes, the show must go on, and so did the big organizations who took the challenge of lockdown as an opportunity and kept running their businesses. Only those organizations could survive these turbulent times that provided the solution of sustaining life and working simultaneously with the aid of technology and opened the possibility of working from home. This initiative was new in most developing countries like India, where the employers with a sceptical mindset allowed remote work, and employees with initial stress and comfort of home agreed to continue to work. The first few months were testing grounds for employers and employees to check on parameters like trust, empathy, insecurity, a relationship beyond work, emotional connection to unhappiness, and strict decisions trying to streamline work. New avenues to work patterns and policies emerged during these times which started railing the derailed business. Most of the organizations were pilot testing their initiatives and their contribution to the productivity of the business and its bottom lines.

The second wave was more critical to the lives of people but the businesses learned and adjusted themselves from the lessons and experiments of the first wave. As time passed and medical experts and researchers rolled out vaccine drives, the restrictions on businesses were reduced by following strict pandemic-related protocols. The third wave of covid is also witnessed by seeing multiple mutations and increased spread of the virus, but this time everyone was proactive and took all due steps to escape the dangers of the new variant. The re-emergence of the covid wave after wave reflects the challenges of the uncertain future and has taught us the power to adjust and adapt to the changing times ahead.

6.1.2 Human Resource Transformation

The HR function captures the entire life cycle of an employee from the day an employee joins and continues to work till the time he/she leaves. HR transformation is the process of fundamentally redesigning and rechartering the HR functions of an organization. In a dynamic world with continuous disruptions, HR issues emerge as a top priority for the top management to thrive businesses for survival, sustenance, and success. Therefore, HR demands major shifts in mindset, roles, capabilities, and digital enablers with reinvention at the core. Enterprises are fundamentally shifting with new business models, technologies, and changing expectations of – and by – the workforce. This creates an unprecedented opportunity for HR to play a new and vital role in shaping the way enterprises compete, access talent, and show up in the communities where they operate. Leaving the traditional mindset of just recruitment, selection, training, compensation, performance appraisal, and industrial relations, the HR function has moved ahead and added new dimensions of working like remote working, employee well-being, reskilling, and upskilling, automation has been initiated. The main objective of HR is to create a happy and productive work culture for its employees so that the business grows multifold:

> The future is an augmented workplace – where humans will focus on value-added jobs and automation will take care of rote processes.
>
> **Poorav Sheth – Chief Digital Officer, Godrej & Bayer**

These new-age practices are in line with the current disruptions in the business environment and prove to enhance the true potential of its employees thereby giving scope to employers to cut down their expenditure and focus on investing in their workforce. As a result, HR has outshined this challenging phase and has evolved as a front-line decision-maker.

6.1.3 Progressive Paradigm Shift

From the very beginning, the rise of mass industrialization has witnessed the expansion and evolution of HR and thus it remains an essential arm of business operations; the evolution of HR as a strategic function has undergone a sea change over several decades (see Figure 6.1).

i. **Manual Revolution** – From the past primeval times and for a continuously long time, goods were produced by skilled artists and craftsmen. Workforce designed and developed the tools and techniques

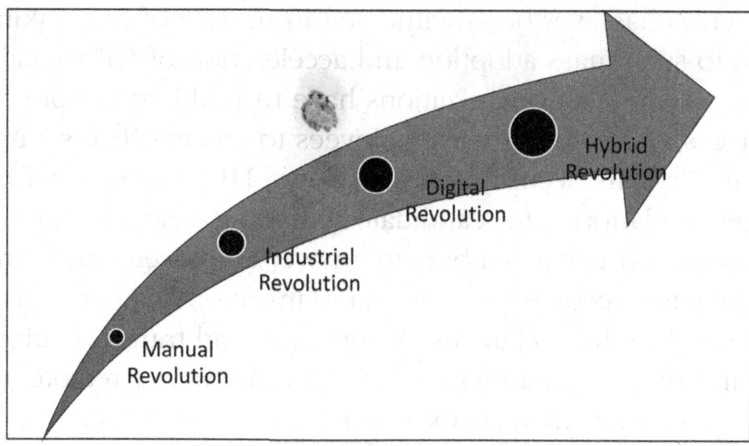

Figure 6.1 Progressive paradigm shift in HR. (Adapted from Gartner's (2020) report "Redesigning Work for Hybrid future".)

to produce the articles and trade them in the open market. There was no employer-employee relation as the family members collectively completed the entire task. The transfer of this expertise was easily transferred from one generation to another.

ii. **Industrial Revolution** – This phase started with the advent of electricity thereby reducing human efforts and enabling fast production. The business owners were concerned about multiplying their profit margins and viciously exploited the human efforts in managing day-to-day work. The workforce in due course of time realized that individually they were powerless and exploited but as united they were strong in the effective functioning of the organization. This realization encouraged them to unite and put pressure on the owners to take productive steps for improving their conditions at work.

iii. **Digital Revolution** – Henceforth, as the time continued and the intricacy of managing employees increased in many business organizations, the purview of industrial relations stretched to different levels of employees. With the invention of the internet and electronic devices like desktops and laptops, work became portable and flexible bringing in global avenues and opportunities. The restriction of place, time, and effort was cornered by virtual platforms and software apps; now the emphasis was on quantity, quality, and efficiency of employees in mainstream tasks.

iv. **Hybrid Revolution** – Worldwide pandemic has led to one of the riskiest experiments in the history of work and the workforce. No similar crisis has ever been witnessed in the era of humankind that led to such mass adoption and acceleration of HR-Digital transformation. Most organizations have turned into remote workplaces and helped their employees to create office set-ups at home. Virtually accustomed workflows; HR analytics; Artificial Intelligence platforms for candidate screening, selection, and onboarding and using chatbots to answer employee issues and concerns have become a reality as organizations have readily accepted and adapted to these changes. Automation and transformation in HR are an instant shift to a blended work mode, where remote work is a new normal rather than an exception.

6.2 Literature Review

HR management mainly focused on traditional paperwork, administrative tasks, and policy framework, despite its strategic role in the organization (Sandholtz et al., 2019). The transition from mundane function to new evolved function in the HR department is happening at a slow pace. However, pandemic lessons have made the HR department realize the significance of the workforce and its transformation for achieving the bottom line for businesses (DiClaudio, 2019). This is evident from the fact that several surveys at the national and international levels reported facts and information on how organizations are transitioning from the usual way of working to the recent new norms. According to a survey, only 17% of companies across the globe use analytics for predicting HR decisions, while 22% report that they are assessing these tools' impact on business results (Sierra-Cedar, 2019). Deloitte's (2019) survey report analysed that 26% of personnel use analytics whereas KPMG (2019) identified that only 20% of HR experts responded to using analytics as a primary HR initiative in the last two years.

According to a survey by KPMG (2020), 60% of HR executives said that HR must adopt a modern approach to address the impending issues of the coming times (see Figure 6.2).

In other words, most of the data highlighted the importance of HR moving to the next level of automation and analytics as a digital revolution, and very soon the traditional approach of HR will cease to exist.

Summary of key findings

Organisation support and well-being

- While **68 per cent** responding organisations admitted that they are **mature to support remote working**, only **48 per cent** of the organisations are supporting their employees by providing **laptops with secured connection** to ensure smooth remote working.
- **72 per cent** respondents confirmed the adherence of basic precautionary health measures like usage of sanitizers by their firms, select organisations have gone a mile ahead to ensure safety of their on-site employees by adopting practices such as **daily fumigation** of transport buses, plant workspace, **boosting immunity** through healthy supplements and food etc.

Employee engagement and communication

- **75 per cent** of organisations have re-defined their communication strategy to increase engagement of employees, **virtual team meetings** and **briefing for employees by leadership** being the top two leading engagement practices.
- Additionally, few companies have enabled **AI-enabled pulse surveys** to capture well- being of employees more frequently.

Learning and development

- Current crisis is pushing majority of corporates to take the leap and switch to **e-webinars (27 per cent)** and **'e-learning' (26 per cent).**

Long-term perspective

- If COVID-19 situation persists, around **22 per cent** of the organisations will defer, freeze or suspend incentive payouts to support their overall finances.
- Few organisations are also exploring to offer **Voluntary Retirement Service (VRS)** to employees.

Compensation and benefits

- While **50 per cent** organisations across industries are keeping their **salary increment budgets unchanged**, **around 36 per cent** organisations have opted for decreasing the salary increment budgets.
- Incentives such as Short Term Incentive (STI), Long Term Incentive (LTI) and sales incentives are being kept unchanged across levels by majority of industries; however **28 per cent** of the responding organisations admitted to having **reduced STIs at the senior and top management levels.**
- With health of employees being the focus for all organisations, there was a positive trend observed in some sectors such as advisory, consumer goods, etc. have reported an upward revision of the insurance benefits.

Promotions

- **50 per cent** of the companies have deferred or suspended their promotion schedule; at the same time a downward trend on promotions numbers across all job levels was observed wherein **33 per cent** of organisations admitted to having reduced it.

Recruitment

- **66 per cent** of organisations have deferred or suspended their hiring schedule at different job levels, while **30 per cent** have also reduced their headcount budgets.
- Contract/part-time/gig workforce are the most impacted by this downward trend in recruitment.

2

KPMG

Figure 6.2 "Future of HR 2020: Which path are you taking?". (From KPMG 2020.)

According to the SHRM (2020) research survey in Figure 6.3, over 7 in 10 employees are struggling to adapt to remote work, 83% of the employers have made business practice adjustments as a result of this pandemic, overall, three-fourths of the employers are having employees working from home. HBR's (2022) report found that 57% of the companies have started using analytics across multiple systems. However, certain problems must be addressed like data collection, data competency skills, capturing behavioural data points, and finding associations between key variables. Multiple minds must collaborate to correlate the HR data with financial data. The datafication of work and workforce leads to high potential in the future and it will completely be a game-changer for future organizations.

According to Gartner (2020), since more employees work remotely, greater is the role of employers as a safety net and wider use of contingent workers, thereby leading to a crisis in critical skills. Dehumanizing efforts, value proposition, culture, resilience, and empathy were factors that were used to deal with employees by HR (see Figure 6.4).

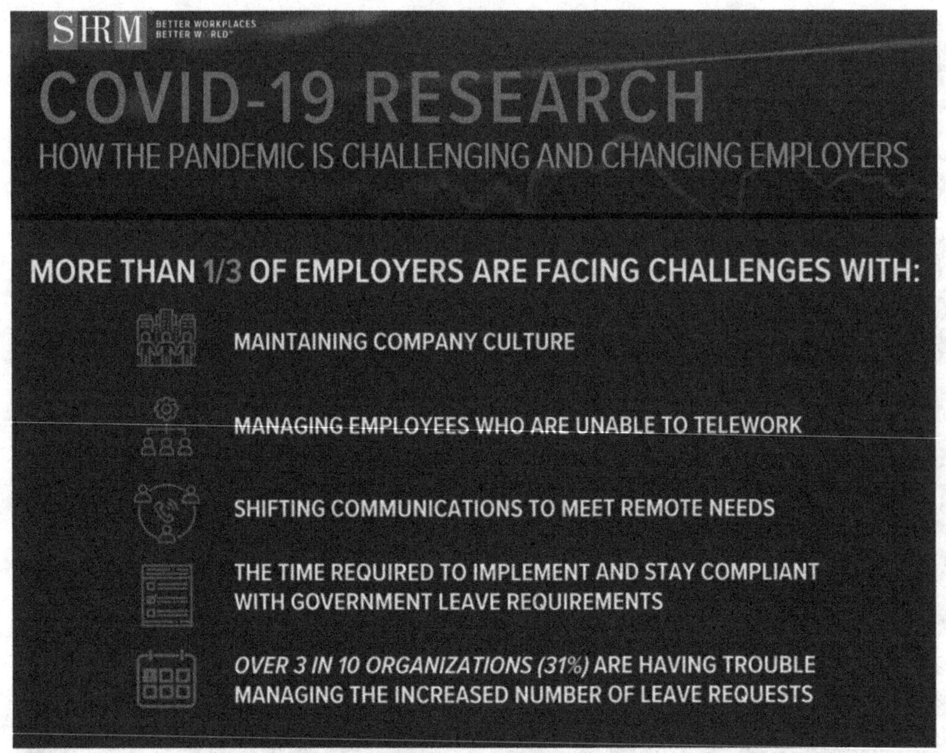

Figure 6.3 "COVID 19-Research: How the pandemic is challenging and changing the employers". (From SHRM, 2020.)

According to another study by PWC (2017), the workforce of the future highlighted technological breakthroughs; demographic shifts in age, gender, work profile, skills, and competencies; a rapid shift from developed to developing countries, and a scarcity of resources and climate changes.

It is imperative from the above studies that the pandemic has increased the pace of the digital revolution and this has impacted the role of HR function as well. The strategic role of HR has taken a new shape whereby dealing with employees for the long-term benefit of the organization has been redefined. Hence HR transformation is the reality for organizations.

6.3 Research Methodology

The main objective of this research is to find out the HR transformation initiatives taken by the organizations in unprecedented times to support the well-being of their employees and in return sustain business operations. A descriptive qualitative-based research technique is used for the current study

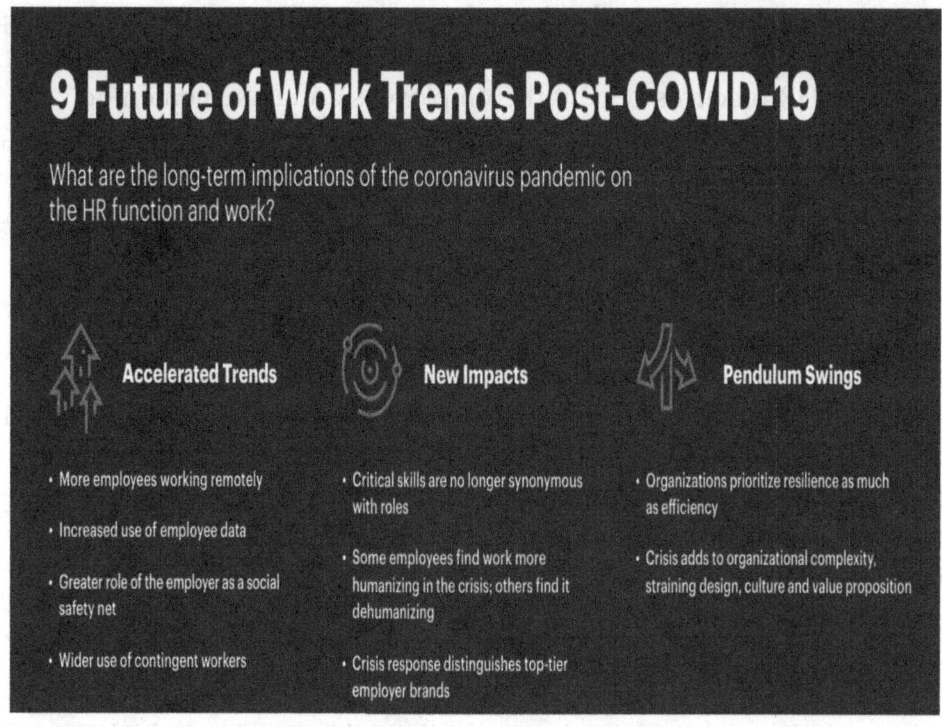

Figure 6.4 "Future of Work Trends – Post-Covid 19, Long term impacts and actions of HR". (From Gartner, 2020.)

based on the primary and secondary data collection. The primary data was collected through interviews with HR managers, executives, senior managers, team leads, etc. through telephonic interactions, WhatsApp calls, Zoom calls, Google forms, LinkedIn responses, face-to-face interviews, etc. Exclusive scanning and scrutiny of survey reports of the national and international research firms like KPMG, Harvard Business Review, Aon Hewitt, SHRM, McKinsey, Deloitte surveys, along with Business Articles, and Webinars was undertaken for secondary data sources. The coding technique using tabulation was adopted to identify common practices for qualitative research.

Primary research responses were low due to the known factors as respondents were working from home, family challenges, and time crunch. Only 40% of responses were received, and only 10% of them were found to be valid and justified to be used for the study. The sample size was 75 HR executives working at different levels among various organizations in the IT/ITeS industry.

The questionnaire itself was the product of the conversations and interactions by HR experts during the test run. The feedback of the experts was given due priority to get actual insights by using subjective questions rather than polished tick mark questions. The qualitative insights were collected and codified to form a broader framework and how these evolutionary HR practices benefit the employees and management to survive, compete, and succeed.

6.4 Data Analysis

The data collected from the respondents along with the surveys and reports by secondary expert sources has resulted in key findings which are quite relevant and well justified. The responses were broadly classified and graphically represented with detailed insights into the initiatives and strategies implemented in these turbulent times. Let us take a look at the analysis of data collected through the interview schedule and secondary sources.

6.4.1 Demographic Profile

The demography of the respondents was limited to gender and job role, wherein males constituted 33% (25) and females were 67% (50). The job level was composed of a maximum of entry-level (80%), then middle-level (16%) and senior-level were (4%) out of the total sample size.

6.4.2 Pandemic as an Opportunity

McKinsey (2020) in their report of 2020 reported that the pandemic has brought a new vision to business by developing new ways of transacting the day-to-day activities. This report also suggested that companies are changing their course as to how they used to operate in the past to survive, sustain, and succeed. Few companies have taken this as an opportunity to expand their scope of business and enter into new horizons altogether.

The graphical presentation (Figure 6.5) represents that almost all the respondents agreed that this pandemic has allowed changing the traditional working process to some new form of work. Even if this change was by force, it initiated everyone to come out of their comfort zone and explore the possible options to perform. At an initial stage, teething problems existed and as soon as the employees got accustomed, they used innovative minds to find new dimensions of work.

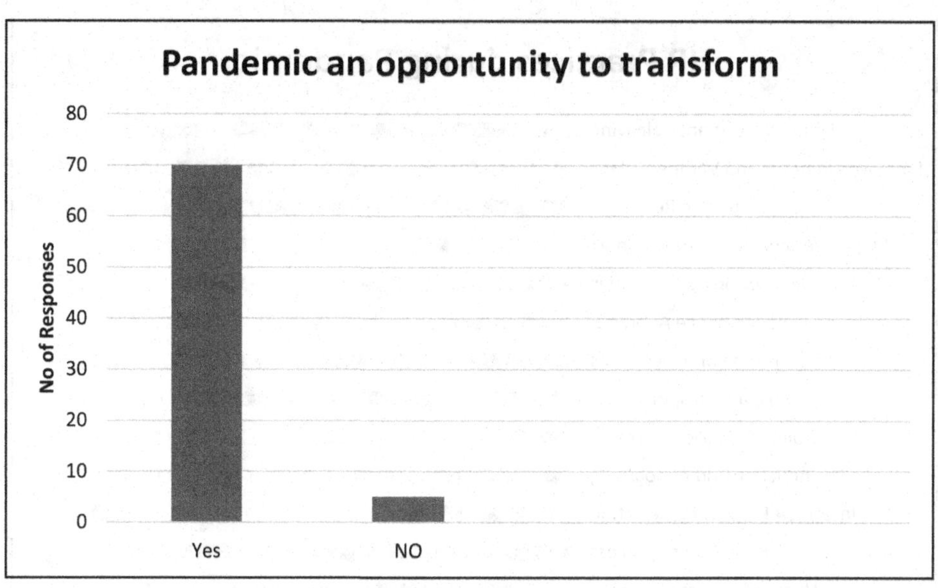

Figure 6.5 Graphical presentation of HR expert responses on pandemic as an opportunity to transform.

The data suggests that the pandemic has been an opportunity to look beyond the normal and be prepared to change when the circumstances ask for it. *Survive and Thrive* are two commonly used terms initiated in this phase and have been the mantra for most multinational organizations.

6.4.3 HR Practices in Pandemic

The pandemic has forced employers and employees to change their prevailing norms of work and come out of their ease zone to explore the different possibilities of work to suit the situation. The responses indicate that most of the HR practices changed to new versions like HR planning, selection, induction, and onboarding, training and development, performance appraisal, employee motivation and support, employee benefits and wellness employee exit and relieving; whereas few practices remained the same as earlier like recruitment process, job role and responsibility, promotion opportunities, and employee retention (see Figure 6.6).

Since this change was unwelcomed and forced everyone to adjust and adapt, the transition to the new HR practices was abrupt; just like the work from home practice. Many HR practices were carried out virtually like the interview process, onboarding of new hires, training for upskilling on key competencies, employee well – a plan which many companies favoured, and

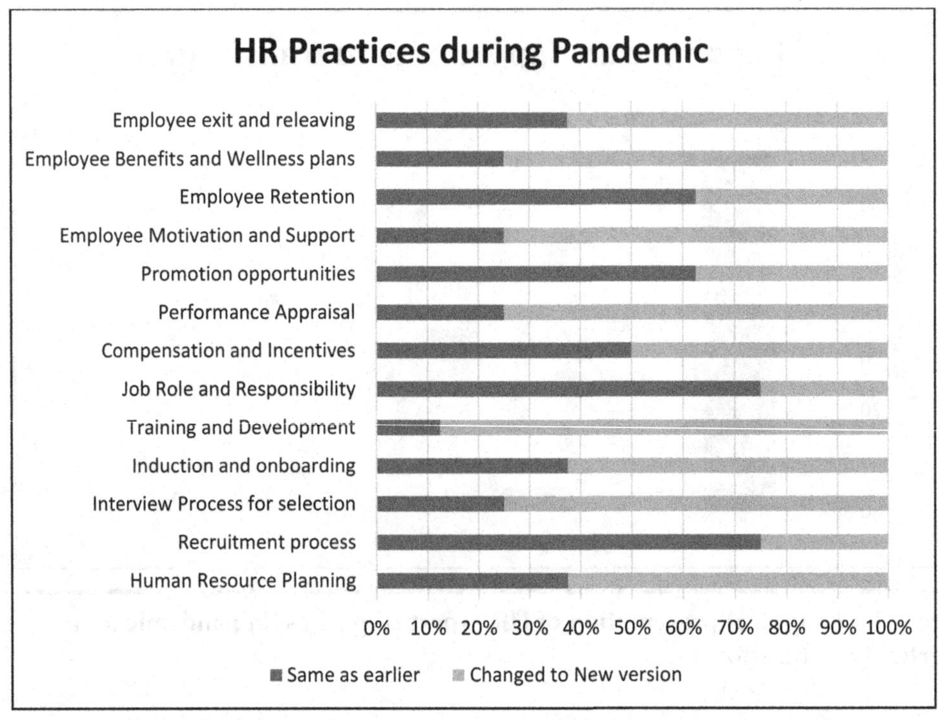

Figure 6.6 Graphical presentation of HR expert responses on HR practices during pandemic.

as a matter of fact, the employees turned out to be more trustworthy and loyal.

6.4.4 HR Transformation

This graphical representation in Figure 6.7 indicates the new dimensions of work that emerged due to covid pandemic across the globe. The responses of the primary survey indicate that there was an instant hybrid work decision which best suited both the employees and employers as the government guidelines strictly forced lockdown. Therefore, physical mode to alternate virtual mode was allowed in most of the job roles. However, there were certain job demands and responsibilities in the area of manufacturing, banking, pharma, medical, etc. where the physical presence of the workforce could not be completely avoided but alternative arrangements like shifts, etc. were made to meet the challenge.

These transformations in less time were made possible due to technological advancement as most of the employees were allowed to

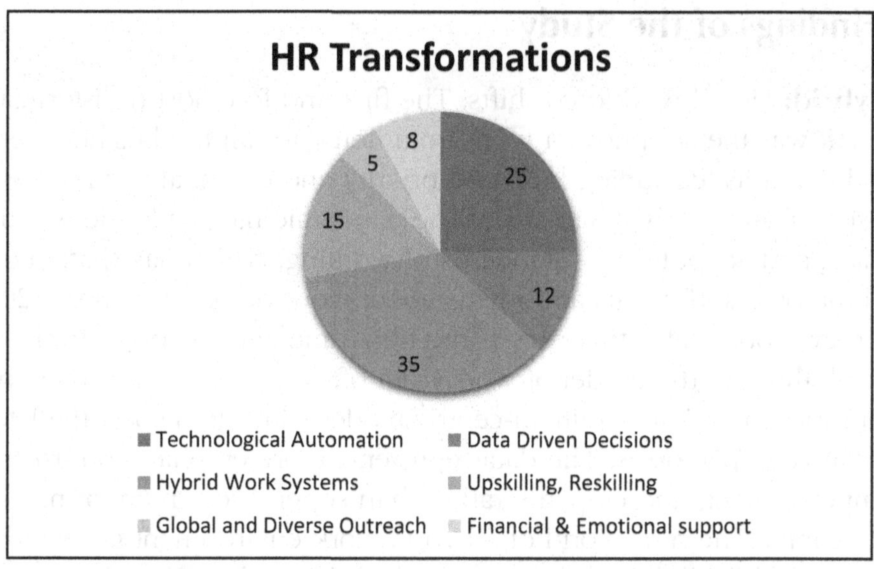

Figure 6.7 Graphical presentations of HR expert responses on HR transformations in work, workforce, and work structure.

work remotely using electronic gadgets through virtual access to perform their tasks. Needless to mention, for the IT/ITeS sector, this phase was not as challenging since work-from-home and virtual meetings are not a new concept for the employees. However, for sectors like education, manufacturing, and banking, it was a daunting task to move processes virtually. Most of the employees accepted the change from physical mode to the virtual model for upskilling and gained expertise.

Earlier in the pre-pandemic era, data-driven decisions were always a daunting task for managers. HR transformation made it possible for all employees to report remotely, which leads to data collection, data collation, and review to draw insights which in turn helped HR to transform into best practices.

The support for the employees was more on the emotional front than the financial which was the need of the hour during the pandemic. The HR practices that focused on empathy for the workforce were put into practice by managers. Another HR practice that came into force was the global and diverse outreach of employees of expertise and high-level skills; this was possible due to virtual platforms as now anyone can work from any corner of the globe. However, this practice of working from home is a temporary phase during a pandemic, but it will bring down the operating cost and generate better-skilled workers in the long run.

6.5 Findings of the Study

i. **Hybrid:** Liquid workforce shifts: The first and foremost transformation in HR was the adoption of work from home for all the business verticals and domains (excluding high-end priority and essential services) as a result of the deadly covid crisis. The pandemic has made the workforce adjust and adapt to new standards of working, which has shattered all the outdated norms and myths about work design. Gartner's (2021) survey report states that now is the right time and the opportunity to break through the model of work from office to work from home by transforming it into a human-centric model with the help of the latest technology platforms. The data represents more emphasis on trust, empowerment, and empathy rather than supervision, monitoring, and directions. The new world of a hybrid work environment gives more access and flexibility to employees which helps them to leverage work and optimize their contribution leading to a win-win situation for both employers as well as the employees.

ii. **Technology-Driven Automation:** The reliance on technology and automation has increased multifold due to the limitation of physical movement. The scope of the digital platforms and apps using artificial intelligence and machine learning to link and grow the work is unlimited. The virtual assistance, digital automation experts, chatbots, and deep learning technology-driven platforms have customized and given the desired results in milliseconds. With the plethora of advantages of the digital value chain, blockchain technology, and automation of back-office administrative work comes the vulnerability of warnings of virus attacks; cyber-attacks; digital crimes and thefts; personal attacks and harassment, etc. Keeping a check on the confidentiality of data with super security checks, firewalls, and advanced antivirus software, business is on the technology-driven highway to reach anytime, anywhere in any form at the speed of light.

iii. **Upskilling, Reskilling, and Lifelong Learning:** The sudden increase of virtual platforms due to remote working has provided diverse opportunities and demands for the employees to unlearn traditional practices and relearn the new parameters of work which expects different skills to support business priorities and challenges. McKinsey's (2021) report has clearly stated that skill-building has been reported to rise among employees (nearly 69% of the companies are emphasizing the skill-building of their employees). The best strategy to develop this is to first assess the gap, then design the skill-building programme and then shift to the new work frame. The digital platforms are in high demand for upskilling as compared to the physical post-pandemic. The experts also confirmed that this culture of learning will continue and develop into the concept of lifelong learning for the employees' development.

iv. **Data-Driven Decisions:** Several surveys and reports concluded that companies that use data analytics for their decision-making mostly perform better. Yet the number of organizations using analytics is very few as the skills and competency required for filling this gap are limited. According to a report published by HBR (2022) many organizations are now expecting to increase their dependency on HR analytics in the next coming years. However, they have initial teething discomfort with data collection, recording, analysis, and interpretation of qualitative data. The challenge of correlating HR outcomes with business results is also crucial. To handle this roadblock, the collective effort of HR and Finance experts must be promoted and in the next coming 2–3 years of testing and trial bright and productive results will be realized.

v. **Employee Well-Being and Benefits:** During the existing situation of uncertain times due to this diverse and dangerous mutation of the corona, there is an increased level of stress, tension, undue emotional pressure, and burnout among employees at all levels of management. This phase provides HR a platform to prioritize human concerns and sensitize the HR practices and policies to meet the needs of the employees in terms of cash and kind incentives. Employee engagement strategies should be reframed to keep the employees motivated and productive.

According to several research reports, the majority of the companies have redefined the work policies to incorporate the existing issues and concerns of the employees. Medical benefit schemes, financial incentives to employees in emergency and need, paid leave facility to covid affected employees, etc. are very common now. Virtual meetings have appeared as the most beneficial form of communication regularly. Organization sensing through pulse surveys, employee connection, and regular interaction by top management provided emotional support to the employees. Many of the MNCs have introduced Yoga, meditation, expert talks, recreational sessions, webinars on impending employee issues, health screening, increasing medical and life insurance contributions, and counselling sessions to keep the employees productive and positive. Even financial support is provided for genuine, emergency cases. All this has created a family atmosphere where concern, loyalty, and empathy exist. Still, there were certain issues and concerns which are not given any possible solution like monotony at work, inefficient employees and work-life balance, family concerns of employees in joint families and with small kids, which must be revisited with the amendments to ensure systematic and healthy work environment.

6.6 Future Scope of the Study

The cloud of uncertainty and volatility is the new normal for the organizations, where future planning can only be done for days to weeks to a few months ahead. Most organizations are focusing on optimizing costs for continuity in business and keeping the employees emotionally and physically stable. Companies severely affected by the covid pandemic are looking for options of saving costs by sending employees on unpaid leaves till the business revives. This protects the organization from unnecessary expenses and provides an opportunity to get new projects or a new line of business. The advancement in technology has provided virtual office space for employees at any location of their choice.

McKinsey (2021) reports that the future of business lies in moving forward towards new levels of work by rethinking the workplace, reforming the culture, digital enablement, upskilling, and reskilling to meet future requirements, and being flexible for an unpredictable future.

The findings of the study clearly defined the benefits of such practices to survive and thrive in this Volatility, Uncertainty, Complexity, and Ambiguity (VUCA) environment. The pandemic has jittered not only businesses but every individual associated with the organizations, many faced losses in terms of deaths of family members, hospitalization due to disease, fear of uncertainties that can lead to emotional distress, financial losses, job change, future insecurity, etc. On the other hand, it also created another mindset in these individuals and organizations, for instance, positive attitude, jugaad, work design, business expansion, resilience, flexibility, and innovation to adapt to the change and move on in life. So, the lessons from this phase have opened up opportunities to put ideas to implementation and create a human-centric work structure. This is not an end, but just the beginning. The technology-driven workforce is just the tip of the iceberg to a digital revolution of the future. The researchers can further explore the possibilities of innovation and reformation in the field of HR during the pandemic and post-pandemic eras.

> As we came out of the pandemic, businesses are not going back to the way things were done beforehand. Digital and People strategy has jumped the queue and is now at the forefront of the C-suite change agenda and it is allowing businesses the freedom to modernize without the red tape of the past – it is simple to transform or die
>
> **Partner of MMB David Dodd**

6.7 Limitations of the Study

The study was conducted in the short period of two years 2020–2022, where the external threat of covid pandemic has affected the entire world at large. Due to the lockdown and restrictions by most of the nations, the employees were less responsive to the questions raised and even delayed their responses. There was again a limitation of not covering all the sectors and industries in the study but the literature review covered almost all the diverse domains. The statistical analysis was not inferential and covered the descriptive framework only. The restraint of time, money, and effort narrowed the scope of the study which can become an opportunity for future research as well.

6.8 Conclusion

This pandemic has clutched the organizations unprepared and left the businesses unsettled in their functioning during the first wave. The HR function which is responsible for the workforce has taken the driver's seat to meet the challenges mushroomed by covid. Needless to mention, the pandemic left the employees with agony and fear due to loss in the family, pay cuts, poor health, uncertainties, and anxieties which have to be dealt with support and empathy by the managers.

As time went by and the increase in the vaccination drive, most organizations started to steer the off-road with the hope of getting back on track with belief in their conviction towards the brighter and greener pastures of work. This chapter underlines the various reformation in the field of HR transformations by reviewing data collected through secondary sources. The primary and secondary data highlighted the brighter side of this pandemic in the workplace despite having grey areas.

Hybrid work systems; automation via technology; upskilling and reskilling for lifelong learning; employee-centric approach to work and data-driven decisions to remove subjective bias were portrayed as the new practices in the field of HR for enhancing the workforce and work practices both. This generation of employees has witnessed the transformation of times, which is part of history and a stepping stone to an unstoppable future. At last, HR has received the priority at the top level of management that it deserved.

References

Deloitte (2019), "Leading the social enterprise: reinvent with a human focus", available at www2.deloitte.com

DiClaudio, M. (2019), "People analytics and the rise of HR: how data, analytics, and emerging technology can transform human resources (HR) into a profit center", Strategic HR Review, Vol. 18, No. 2, 42–46. 10.1108/SHR-11-2018-0096

Gartner (2020), "Future of work trends – Post-Covid 19, long term impacts and actions of HR", available at https://www.gartner.com/en/human-resources/trends/future-of-work-trends-post-covid-19

HBR (2022), "Empowering the new decision makers to act with modern self-service analytics", available at https://hbr.org/sponsored/2022/01/empowering-the-new-decision-makers-to-act-with-modern-self-service-analytics

KPMG (2019), "The future of HR 2019: in the know or the know the Gulf between action and inertia", available at https://advisory.kpmg.us/content/dam/advisory/en/pdfs/hr-survey-2019- key-findings.pdf

KPMG (2020), "Future of HR 2020: which path are you taking?", available at https://assets.kpmg

McKinsey (2020), "The new science of talent: from roles to returns", McKinsey, available at www. mckinsey.com

McKinsey (2021), "Building workforce skills at scale to thrive during—and after—the COVID-19 crisis", available at https://www.mckinsey.com/business-functions/people-and-organizational-performance/our-insights/building-workforce-skills-at-scale-to-thrive-during-and-after-the-covid-19-crisis

PWC (2017), "Workforce of the future – The competing forces shaping 2030", available at https://www.pwc.com/gx/en/services/people-organisation/publications/workforce-of-the-future.html

Sandholtz, K., Chung, D. and Waisberg, I. (2019), "The double-edged sword of jurisdictional entrenchment: explaining human resources professionals' failed strategic repositioning", Organization Science, Vol. 30, No. 6, pp. 1349–1367.

SHRM (2020), "Covid 19-Research: How the pandemic is challenging and changing the employers", available at https://shrm.org/hr-today/trends-and-forecasting/research-and surveys/Documents/SHRM%20CV19%20Research%20Presentation%20Release%202.pdf

Sierra-Cedar (2019), "The sierra-cedar 2019-2020 HR systems survey white paper", 22nd Annual Edition, available at www.sierra-cedar.com

Chapter 7

Workforce Shift after COVID-19 Outbreak: Adapting to Evolving Workplace and Work-Life Settings

Navjot Kaur

MoneyCom, Queensland, Australia; Palliative Care, Queensland, Australia

Contents

DOI: 10.4324/9781003267751-10

7.1 Introduction

The unprecedented advent of COVID-19 has impacted economic activity, employment levels, and systems, causing social and community uneasiness (Cotofan et al. 2021). Not only have the far-reaching consequences of COVID-19 potentially impacted the social well-being of workers but have also dismantled and changed the ways we undertake tasks and perform activities thus changing the world of work. Workplaces have to make significant alterations and use multiple interventions to prepare for workplace disasters by changing policy and strategizing to stabilize job security when faced with public health crisis (Pacheco et al. 2020). Business operations of many organizations went virtual as workplaces were susceptible to making substantial transformations. Extensive research had been undertaken to scrutinize the evolving digital transformation and impact of virtual operations on productivity and corporate culture (Kim 2020).

In January 2020, the COVID-19 pandemic was detected in Australia with an exponential increase in infections (Munawar et al. 2021). Certainly, restricting mobility was the central aim of preventing COVID-19 transmission to lower the contact rates that had enabled the governments and policymakers to make evidence-based decisions (Kim 2020; Khajuria et al. 2021). For instance, the Australian government had implemented domestic control measures such as cancellation of mass gathering activities (nationwide lockdowns), suspension of public transportation, social distancing, quarantine, shutting down of educational and training institutes, and closure of domestic and international airports (Munawar et al. 2021). Consequently, these alterations have significantly affected the work-life and lifestyle of citizens, therefore, influencing their behavioral and social activity patterns.

The graph in Figure 7.1 indicates the searing increase of confirmed COVID-19 cases in March and July 2020. In Victoria, the spike in the number of new infections further elevated the peaks with the highest number of cases (i.e., 721) as reported on 30 July. New South Wales struggled with slowing down the infection pace with nearly 5,000 cases whereas Queensland (nearly 2,000), Western Australia (nearly 2,000), and South Australia (almost 1,500) kept it safe with the lowest transmission rates. Although the crisis is unsettling with new mutants and variants of coronavirus succeeding the previous versions of the disease. By the time this chapter has been jotted down, there are 23,121 new cases in the state of New South Wales, 5,622 cases in Queensland, and 17,636 locally acquired

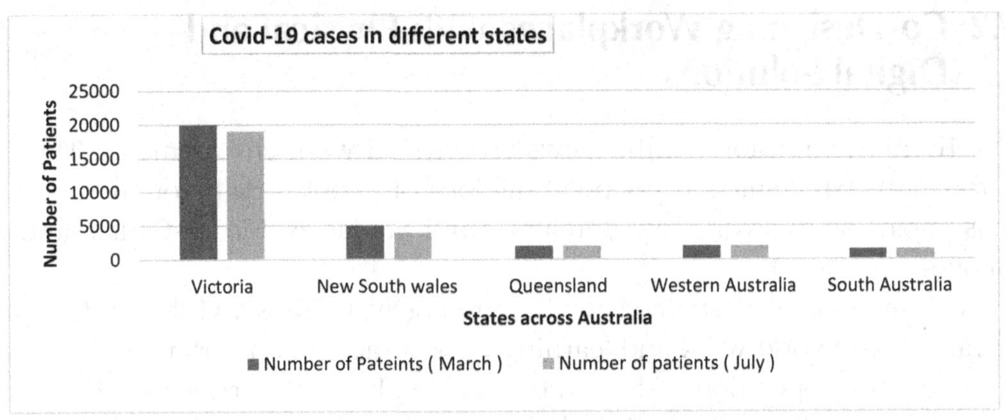

Figure 7.1 Graph indicating several COVID-19 cases across the Australian States reported in 2020 for March and July (Munawar et al. 2021).

new cases recorded in the state of Victoria (Nguyen 2021; Queensland Government 2022; Victorian Government 2022).

Looking at the unsettling statistics, citizens in various states and cities were asked to refrain from social gatherings, limit out-of-home movements, and eventually, businesses have been disposed toward transitioning to remote or virtual workplaces with an expected surge in the layoff of employees. Consequently, the public health orders had imposed restrictions that changed the workplace policies thus making the employees quickly adjust to the digital transformation (Kim 2020). Employees were unable to attend their place of work, especially those who are carers to their young children (unvaccinated) at home. Therefore, the COVID-19 crisis has been dealt with as a matter of urgency, and the response measures are predominantly inclined toward process optimization and product digitization in workplaces (Kim 2020). As a result, the trend of digitalization was anticipated to replace the human gatherings in corporate culture along with flexible working hours (Kim 2020; Cotofan et al. 2021). This enabled the CEOs and managers to predict some long-lasting impacts of the coronavirus pandemic on places of work. Although small businesses such as cafes and other local stores are facing economic turmoil, on the other hand, medium-sized businesses are experimenting with new software and decentralized decision-making to adopt and implement cybernated corporate climate to enhance work productivity and effectiveness in physical workplaces (Kim 2020; Cotofan et al. 2021). However, it is critical for businesses to make this adaptation of hybridization of workplaces to make the workability as efficient and swift as possible (Kim 2020).

7.2 Co-Designing Workplaces with Physical and Digital Solutions

COVID-19 has transformed the views on physical work environments. The coronavirus pandemic has upended millions of people's daily routines, thus impacting the world's work realm (Sandoval-Reyes, Idrovo-Carlier, and Duque-Oliva 2021).

With millions of Australians working remotely, there is a visible shift toward more hybrid work and learning environments even for more incumbents of operations. Also, with increased border restrictions and extended periods of hibernation, future places to engage, explore, and work are now hybridized into a digitally and physically integrated working environment (Yang, Kim, and Hong 2021). Certainly, the usability of the workplace environment depends on the distinct service design, delivery, and office layout methods. Therefore, one of the few silver linings of the COVID-19 was an elevated and improved perceived work-life balance as compared to the non-pandemic era (Yang, Kim, and Hong 2021). To tackle an unprecedented crisis, countries around the globe with limited health-care resources and safety provisions have formally approved and implemented certain constraints which further led to companies and service providers altering the ways they worked. Australia was no exception. The Australian Institute of family studies (AIFS) reported that two-thirds of the employed individuals have been able to work from home (WHF) thus reducing the cost and tedium of commutes (AIFS, Australian Government 2021). Regardless of geographic location and demographic fluctuations across metro and regional areas, transparent and inclusive technological options provided quick and considerable accessibility to work-related information and services. By now, most companies are accelerating their digital capabilities to keep pace, protect employees as well as serve the customers who are facing mobility restrictions due to the pandemic (Baig et al. 2020). These resources were underscored and underutilized before the Covid crisis (Harguth 2021). Digital solutions have increased connectivity and communication through online platforms such as Zoom, Teams, Mural, etc. Due to the global pandemic, digital solutions have become the predominant means of providing up-to-the-minute information thus increasing the regulation and retrieval of economic and social utilities, amidst the period of isolation or lockdown (Harguth 2021).

The combined pillars of digital and remote workplace along with the availability of physical space had become a 'place' in the post-Covid era. Recent data as collated from global consumer surveys show that organizations have vaulted business and consumer e-commerce adoption in a matter of almost two months which was targeted instead for five years as a long-term strategic plan (Baig et al. 2020). For instance, banking businesses in Australia have launched digital outreach programs such as 'Anywhere Operational models' to assist and accelerate the level of remote sales and services which were incomparably low as compared to metropolitan service models (National Australia Bank 2021). Results of the Accenture Future of Work Study 2021 indicated that around 83% of employees prefer a blended or hybrid model which creates a productive Work-from-anywhere (WFA) culture but still leaves a challenge for the employers to authorize and manage the levels through digital decentralization (National Australia Bank 2021). COVID-19 had 'radically impacted' how employees work and adapt in different ways throughout the pandemic. Statistics provided by one of the prominent banks in Australia show that most of the workforce is based in two jurisdictions, i.e., Victoria and New South Wales had experienced 95% of staff working from home in 2021 (National Australia Bank 2021). Similarly, Grocery stores have switched the delivery and ordering through online sources to run their primary business (Baig et al. 2020). Schools and educational organizations in many local areas across Australia have pivoted to 100% online (digital) classrooms for elevating systematized and structured learning and training (Baig et al. 2020). As the list goes on, doctors have begun delivering services through telemedicine or telephonic appointments for the patients which had aided more flexible regulation of the health systems across various areas, i.e., metro, regional, rural, and remote (Baig et al. 2020). WFA provided a choice to the employees to live and work from the location of choice within a specific country with a reliable internet connection. Companies such as Akamai and SAP have developed WFA programs as a consideration to allow employees to WFA (Choudhury, Larson, and Foroughi 2019)

Digitally, there are plenty of IT tools that have come into existence and are much available for professionals in various fields to make their life easier as they work from dispersed locations. To begin with Zoom, Microsoft Teams, Google meets, Hangouts, Google Docs, and Skype were used for streamlining online interactions without much hassle across places and even around the globe (Singh, Kumar, and Ahmad 2020). As reported by Google Meet, 'Zoom' (an online video conference software) had reported almost 60% growth in user activities and transactions with people devoting over 2 billion minutes to virtual meetings every day thus leading to an approximately 78% increase in profit of this digital platform (Kim 2020).

Technically, the industries might have undertaken change management with the deployment of resources to confront and counter unprecedented change, but on top of all this, it is critical to be culturally, psychologically, and emotionally ready for the post-pandemic era (Singh, Kumar, and Ahmad 2020; Cook et al. 2021). However, the potential gains could be counterbalanced by the sociocultural and behavioral challenges which are potentially associated with the present tendency to adapt the enforced provision of usage of virtual and teleworking on a massive scale.

7.3 Organizational Trends and Workplace Dynamics

The existential peril of the coronavirus pandemic has originated a consequential rethink in organizational life and has profoundly transformed organizational cultures (Spicer 2020). Almost universally, organizations have been forced into rapid and radical transformation of programs and processes (Amis and Greenwood 2020). Further transformation to virtual systems in most cases might bring major changes ahead for many organizations that are keen to survive and thrive (Amis and Greenwood 2020). The advent of new features in the technology has started to develop new etiquettes and cultural norms around how training leaders and managers run virtual teams and how work is undertaken and performed collectively (National Australia Bank 2021). With dispersed workforce settings and technology playing a crucial role in supporting and maintaining the workplace systems intact, fostering a supportive work culture is as critical. Inclusivity, leadership, and workplace democracy had taken on new dimensions in a remote-first, virtual-team work, and digital-first environment (Spicer 2020; National Australia Bank 2021). One of the challenges faced by the organizations for people working both on-site and off-site was to ensure the environment of inclusive meetings and collaborative sessions for virtual and remotely located teams. Hence, it is just more than a technological setup for collaborating with the teams for planning, developing, and implementing processes in the workplace. Therefore, as a new inclusion into the work culture of IT, determining the nature of inclusivity and fairness was mandatory as it accounts for behavior, norms, and rituals needed to be established for the virtual work environment and ergonomics.

Symbolic icons of organizational culture as recognized by open-plan workspaces, staff dressed up in suits, and business attire have been substituted by acrylic monitors and other 'personal protective equipment'

(PPE) which included mandatory use of masks not only outdoors but also indoors (Spicer 2020). Staff workplace rituals such as toolbox chats over a cup of coffee and water cooler chats have been replaced by zoom calls. In a research on a call centre, employees elucidated the fact that flexible and remote working has increased their productivity by almost 13% due to a reduction in sick days, recess hours, work-related stress, and sleeping troubles as compared to their office-based counterparts (Choudhury, Larson, and Foroughi 2019). Additionally, a report by the Australian Productivity Commission (APC) highlighted that employees enjoyed benefits such as reduced commute time and travel costs and wanted to WFH (Australian Government 2021). Also, the APC research indicated that a sudden and widespread switch to remote working conditions had not considerably impacted employees' productivity (Australian Government 2021). Statistically, if the case of widespread remote working continues, the level of productivity could increase in direct proportion to the flexibility in employee work time, patterns, and locations (Australian Government 2021). Furthermore, employers can increase the value of employees by granting geographic autonomy and flexibility to work. In contrast, employers were concerned about the possible negative impacts on workplace culture and teamwork. Further, the unprecedented conditions have posed a major challenge to the managers as the fundamental values and presumptions of various businesses have switched from scrutiny, investigation, and ingenuity to health, safety, and adaptability in the workplace (Choudhury, Larson, and Foroughi 2019; Spicer 2020). The research highlighted that an average employee was acquiescent to accept 8% less pay if provided an alternative to WFH thus indicating that workers prefer flexibility over the assigned monetary value provided (Choudhury, Larson, and Foroughi 2019). Exploratory research indicated that during and post-COVID-19 era, the workforce will be entering into a radically new arena where most houses will have a mini-office as an alternative work area and organizations might promote WFH (Singh, Kumar, and Ahmad 2020). Also, findings by Global Research firm Gartner show that 74% of chief financial officers were implementing plans to permanently shift and transition their employees to remote working (or WFH) settings to economize and save the cost for the company (Parungao 2020; Singh, Kumar, and Ahmad 2020).

Although an organization is steered and geared by vision, mission, and values, organizational culture could be demonstrated and validated throughout the course and after times of pandemic, crisis to understand what works well and learn lessons for the future. Moreover, research shows that money-spinning companies exhibit a strong value-driven culture that is people-centric, accountable, collaborative, agile, and works with integrity, innovation, and ambition (Berman and Thurkow 2020). Hence, organizations must recognize the importance and significance of promoting a flexible workplace culture through incorporating endurance, conviction, and teamwork which is further reinforced by vigorous technology to assure sustainability, rationality, and stability in dynamic and challenging situations (Singh, Kumar, and Ahmad 2020).

7.4 Work-Life Integration in Pandemic

Besides being a health extremity and economic hazard, the pandemic has caused considerable disruption of people's roles and responsibilities within communities and households. As mentioned above, to mitigate and control the outspread of the virus, many countries have taken drastic initiatives like physical distancing, curfews, lockdowns, school closures, the closing of public establishments, and other workplaces (Hjálmsdóttir and Bjarnadóttir 2021). A survey conducted by 'The Conversation' indicated that many people felt fortunate to still be a part of the workforce, being in work and fear of being under surveillance due to the collapse of work-life boundaries have led most people to work harder and for longer periods of times (Cook et al. 2021). On the contrary, UNESCO (2020) has estimated that almost 1.3 billion (i.e. more than 70%) students across the globe were affected by school closures, restricted mobility and services thus increasing the responsibilities of the parents/guardians or carers at home (Cook et al. 2021; Hjálmsdóttir and Bjarnadóttir 2021). Those who were juggling work and caring responsibilities struggled the most (Cook et al. 2021). In Australia, more than 7,000 Australians were surveyed by 'Life during COVID-19' to understand how people are adjusting to the social distancing measures, unprecedented lockdowns, and restrictions (Baxter et al. 2020). The most significant changes in the employment status were reported by the households, with around 60% of the respondents indicating – 'always working from home' in contrast to only 7% pre-pandemic employment conditions (Baxter et al. 2020; AIFS, Australian Government 2021). Considerably, it is a hard time for that managing childcare and caring for older relatives at home along with juggling work-life responsibilities. For instance, a UK poll showed that 71% of working mothers disapproved of sabbatical to look after their young ones during the pandemic which further indicates the need for concrete action by the policymakers to protect workers' rights to self-care and break-time when required (Cook et al. 2021). Undoubtedly, households with dual-earning have increased over the years, yet various researches stated that still the females assume the burden of household labor and childrearing in both developed and developing countries (Baxter et al. 2020; Hjálmsdóttir and Bjarnadóttir 2021).

According to the reports by AIFS, 7 in 10 working parents reported that they are 'actively' or 'passively' taking care of the children, but as mentioned above women were five times more likely to perform the role of primary caretaker in the family (Baxter et al. 2020). Studies and media coverage indicated the impact of the pandemic as families face barriers and obstacles along with exaggerated or revealed gender inequalities, power differences, and a divide in household workload (Hjálmsdóttir and Bjarnadóttir 2021). Research shows an unequal and irregular labor allocation in the Australian households as females bear the burden of household chores, childcare as well as emotional labor with distant work pressure and mental work (Baxter et al. 2020). Hence, decision-makers and authorized personnel must aim to narrow the power and gender divide in both public, as well as private organizational settings with a focus on ultimately closing the gender gap, in the homes during the change in circumstances (Baxter et al. 2020; Hjálmsdóttir and Bjarnadóttir 2021). Correspondingly, the 'Covid-19 employer study' revealed that younger employees indicated their inclination and likeliness to work in a formal office environment as compared to older workers or parents who were conveniently adapted to the technologies and gadgets that enabled them to work remotely (Booth 2021). The findings of this study were based on in-depth interviews with 28 organizations (comprised of industrial sectors including retail, transport, advertising, government, and insurance) and 32 leaders and 70% of the participants from large organizations across seven states and territories in Australia (Booth 2021). Undoubtedly, most employers navigated the uncertainty and adapted to flexible and agile work environments that suited the individual staff needs thus moving along with the dramatic shift brought in by the pandemic in the relationship between organizations and their employees (Booth 2021).

Economically, the implications for the organizations during a public health crisis are to particularly enhance the self-efficacy of the staff through managers and team leaders providing effective counseling to determine achievable and realistic goals or targets following well-defined KSAs, i.e., knowledge, skills, and abilities (Nilasari, Nisfiannoor, and Devinta 2021). Additionally, managers need to consider the essential sociopsychological support through work-life balance by assigning clear, detailed, and measurable tasks to employees without any gender or age bias. Various forms of motivation can assist in narrowing the gap of work imbalance in the professional and personal lives of employees. For example, paying equal remuneration for equal work, creating attractive incentives, bonus or commission programs, public recognition of tasks accomplished, and achievements to retain the talent in the challenging times when voluntary termination is high during COVID-19.

7.5 Gendered Impact of COVID-19 Outbreaks

The discussion so far indicated that the COVID-19 crisis is likely to affect women and men differently and might place a disproportionate burden on females, therefore, leading to long-term repercussions for gender equality (Alon et al. 2020). Research and experience from the past outbreaks elucidate the importance of promoting gender and health equities by incorporating gender interpretation and investigation into evaluation and response to ameliorate the implementation of health and workplace safety interventions (Wenham, Smith, and Morgan 2020). Unlike previously experienced recessions, COVID-19 is much more likely to reduce employment in service sectors where women form a large fraction of the workforce (Alon et al. 2020). Research shows that the pandemic has posed a challenge for women to advance in their careers as more women (as compared to men) have lost their jobs and women in essential services are more exposed to infections and psychological stressors (Carli 2020). For instance, considering their front-line interaction with the clients and communities in the workplace could be considered a concerning factor that women have not been completely incorporated into global health security surveillance, prevention, and detection mechanisms in specified workplaces (Wenham, Smith, and Morgan 2020). Perhaps even more importantly, females have encountered more work disruption than males because of the increase in childcare needs and other responsibilities that stem from the closing of schools and day care facilities across Australia (Carli 2020; Wenham, Smith, and Morgan 2020). Undoubtedly, single mothers are more severely impacted as they are proportionally outnumber single fathers by a large margin (Alon et al. 2020). A study conducted by Lancet shows that during the outbreak, men are more likely than women to have authority in decision-making leading to considerably high unmet social and financial needs among women (Wenham, Smith, and Morgan 2020). In contrast, telecommunication has also increased men's responsibilities for childcare thereby narrowing down the gender gap in the long run in domestic responsibilities, therefore, increasing gender equality to a certain extent (Carli 2020).

Sadly, lockdowns and WFH have also led to a rise in the incidence within the families due to severity in relationships, domestic violence, and family violence against women (Kinnersly 2021). Undoubtedly, Australian workplaces must take cognizance of those who encounter various kinds of direct or indirect biases such as ageism, racism, xenophobia, homophobia, and ableism along with the ongoing impact of colonization that can broaden the gender inequality (Kinnersly 2021). As these findings are concerning, rigid gender stereotypes have led to the brunt of the rise of unpaid care demands for women (as well as men in some cases) with so many struggling to manage home-schooling, restricted childcare access with WFA alternatives (Carli 2020; Wenham, Smith, and Morgan 2020; Kinnersly 2021). In May 2021, the Australian Bureau of Statistics's 'Household Impacts of Covid survey' established that 'females expended 20 or more hours a week on unpaid care, homeschooling, and supervision of children' as compared to household duties undertaken by men at home (Kinnersly 2021). Consequently, these expectations can reduce the financial freedom and independence of females which can further devalue their worth in the societal and economic setup.

A research study undertaken by Mckinsey and Co. highlighted that, if policymakers and business leaders modeled potential do-nothing scenarios to sustain and proceed with gender discrimination, it could adversely impact the social and economic lives of females (Madgavkar et al. 2020). Therefore, to reduce the economic downturn due to gender inequality in the workplace, investing in girls and women can represent a significant opportunity to drive inclusive economic growth and gender equity (Madgavkar et al. 2020). Therefore, to eliminate the perpetuation or reproduction of gender and health inequities, organizations may directly or indirectly analyze the gendered impact and incorporate voices of women within preparedness and response policies or workplace practices going forward pre and post-outbreak (Wenham, Smith, and Morgan 2020).

7.6 The Resilience of Employees during the Pandemic

An organization's resilience has become indispensable to business continuity and recovery during the time of the Covid crisis. Nonetheless, the possibility of a resurgence of new strains of COVID-19 had compelled communities and organizations to adapt to the Covid-prone environment as new normal. As the infection curve flattens, many regions across Australia have started easing restrictions on lockdowns and homestays, and workers were allowed to share office spaces (Ojo, Fawehinmi, and Yusliza 2021). Australian workers maintained a relatively stable sense of well-being despite the elevated levels of stress and anxiety throughout the pandemic (Lycett 2021).

The 2020 survey conducted during the lockdowns shows that most Australians (96%) reported having greater empathy for others and almost as many (96%) indicated having more gratitude for the things they have in their life (Lycett 2021). Research by Deakin University indicated that, whilst many Australians reported positives it also remains a challenging time for many vulnerable people due to pre-existing factors, potentially compounded by a pandemic or consequently exaggerated due to pandemic (Lycett 2021). However, the pandemic as a dreadful episode has been exceptionally demanding for caretakers and parents (Mikocka-Walus et al. 2021). Federal and state restrictions in Australia included authorized working from home, remote learning from home for college and university students, limited social activities, and shutdown of community centres, playgrounds, and sports hubs (Mikocka-Walus et al. 2021). Consequently, kids were restricted from any participation in outdoor activities, hence parents and caretakers were bound to undertake exceptional accountability for kid's home-schooling and managing affairs of extra-curricular activities around social and welfare needs while undertaking extra tasks on as working from home (Mikocka-Walus et al. 2021). Research shows that almost 66% of the parents stated that they were inept to provide for all their requirements and meet the twofold demands of workplace and their kid's well-being during the pandemic (Mikocka-Walus et al. 2021). Most importantly, research demonstrated that loneliness was one of the key factors that contributed to the distress, over and above therefore influencing the levels of 'psychological resilience' (Mikocka-Walus et al. 2021).

7.7 Leadership and Team Management during Crisis

The pandemic has taken a humanitarian toll by placing extraordinary demands on leaders and fear among employees and other stakeholders in business and beyond. Leaders in organizations have to manage elements such as volatility, uncertainty, ambiguity, and complexity that needed rapid, high-impact decision-making with limited information and within traditional organizational settings (Kaul, Shah, and El-Serag 2020). Leaders, managers, staff, and stakeholders must learn new lessons along the way for innovative and creative ideas and problem-solving strategies to continue the business and enterprises up and running. Although a by-product of uncertainty is anxiety which swept throughout the workforce and leadership was silenced during the confusing times (Kaul, Shah, and El-Serag 2020). During times of turmoil, clear, consistent, and adaptive communication by the leaders played a key role to sustain the organizations (Kaul, Shah, and El-Serag 2020). Undoubtedly, there was a desperate need of leveraging the team culture and leadership with an ability to communicate credible hope about available resources that meet the needs to tackle the threats during health and economic crisis (Forster, Patlas, and Lexa 2020; Raghuram 2021). In addition to healthy human resources to take on the charge of work, both material (i.e., PPE, health and safety resources, mental wellness resources and technical resources, etc.) and psychological resources (i.e., determination, solidarity, shared purpose, and shared responsibility, etc.) play a key role in crisis management along with an ability to work remotely (Forster, Patlas, and Lexa 2020; Kaul, Shah, and El-Serag 2020)

Research shows that success under crisis is determined by the leader's caliber and competency to process information, behave precisely, and influence others both internal and external to the organization (Lagowska, Sobral, and Furtado 2020). For instance, instead of acting impulsively or freezing under situational pressure, leaders should be capable of generating reasonable solutions, credible options, and clear criteria to evaluate the actions and plans for effective decision-making (Lagowska, Sobral, and Furtado 2020). Though these abilities are fundamental to leadership capability irrespective of the situations but to perform effectively, leaders need to stay unbiased and unprejudiced to make required adjustments for an exceptionally well response during the crisis by being more adaptive (Kaul, Shah, and El-Serag 2020; Lagowska, Sobral, and Furtado 2020). With a high degree of precariousness and uncertainty, decision-making for future developments succeeding in the mitigation of crisis could prove as an effective competency check for leaders in the organization (Lagowska, Sobral, and Furtado 2020). Furthermore, effective processes and productivity could be achieved through mindfulness about individual biases while collecting, assessing, and analyzing multiple fragments of data (or information) from various reliable sources (Hadley et al. 2011)

The unprecedented situation due to COVID-19 had radically transformed the manner people think, perceive, work, converse, and mingle. Also, this situation emphasized various reflections on the role and responsibilities of a leader to manage remote employees during changed circumstances (Lagowska, Sobral, and Furtado 2020). For example, stepping out of the traditional setup of the workplace and leadership behaviors could be a confusing experience, not only for the teams but also for the team leaders (Cho 2020; Lagowska, Sobral, and Furtado 2020). This can exaggerate issues with task accomplishment, performance management, productivity analysis, appraisals, and employee engagement (Forster, Patlas, and Lexa 2020; Lagowska, Sobral, and Furtado 2020). While most of the research available addresses the role of leaders and practice of leadership in 'Normal context' or 'Pre-Covid era', the same theoretical and practical elements might no longer fit into the context of 'crisis era' or 'during or post-covid era' due to increased uncertainty and elevated disruption in the routines (both personal and professional) of employees (Cho 2020). Although the advent of crisis has increased the scope of research with the provision of the new platform to explore the impact and effects of the crisis, also new studies focused on virtual leadership and crisis management in teams are in pipeline (Lagowska, Sobral, and Furtado 2020).

7.8 Change in Motivation, Perception, and Resilience during the Pandemic

Findings from the pre-pandemic era highlighted that personality is an important individual factor and predictor of performance and optimal delivery of outcomes in the workplace (Barrick 2005). Strong evidence suggests that motivation mediates between personality and performance of individuals thus affecting the behaviour through performance motivation to achieve set goals, expectations, and self-efficacy in the organization (Barrick 2005). Correspondingly, employees driven by capacity, autonomy, and competence are more likely to acquire and develop intrinsic motivation (Nilasari, Nisfiannoor, and Devinta 2021). Intrinsic motivation could be defined as an undertaking or performing a task or activity for inherent satisfaction, enjoyment, and interest rather than following an order for separable consequence (Ryan and Deci 2000). On the other hand, doing something to achieve separable results may or may not fulfill the intrinsic satisfaction of an individual, but tasks performed for monetary goals may refer to extrinsic motivation (Ryan and Deci 2000). Furthermore, multiple studies indicated that self-efficacy and performance are mutually correlated and impact workability (Nilasari, Nisfiannoor, and Devinta 2021).

In the current pandemic-prone working environment, retaining employee motivation has become a human resource (HR) issue. Organizations are planning and implementing different ideologies to develop, maintain, and sustain their HR strategies for high performance, with optimal and timely results both in the short and long term (Nilasari, Nisfiannoor, and Devinta, 2021). A multidimensional conceptualization of motivation and its effect during the Covid revealed that intrinsic motivation during Covid might not increase employee performance (Nilasari, Nisfiannoor, and Devinta 2021). In contrast, extrinsic motivation in the job such as financial possession and social recognition could assist the organizations to retain talent and increase self-efficacy which in turn leads to increased employee performance (Nilasari, Nisfiannoor, and Devinta 2021). Research conducted by the Australian Psychological Society shows that employee engagement could be associated to several outcomes, like job performance, job contentment, commitment to the job role, turnover intentions, stress, burnout, and organizational citizenship behaviours (Australian Psychological Society 2020). Motivated and engaged employees are less likely to make mistakes, provide a high level of service and will, and experience greater levels of subjective well-being (Australian Psychological Society 2020). Therefore, it is important to focus on the factors that drive motivation and engagement, particularly increasing employee resources and well-being. Many aspects such as administrative hassles, organizational policies, grievance, conflict (including role conflict), lack of resources, and an excessive workload can negatively impact the employees, especially when extrinsically they are dealing with a pandemic crisis (Australian Psychological Society 2020; Demirović Bajrami et al. 2021). While it is important to watch out for these negative triggers, it is of significant value to assess the set of circumstances of each employee and to encourage respite from work, ensuring adequate recovery and revival time (Australian Psychological Society 2020). Studies indicated that organizational changes (during and post-Covid) condensed the level of performance, job contentment, job stimulus, and engagement for ones who continued at work as office-based counterparts (such as employees in essential health-care and retail services), and those alterations were perceived as inequitable and reflected negligence of the administrators (Demirović Bajrami et al. 2021). Additionally, the shift to flexible work has been dramatic as employees working remotely were likely to have more screen time thus increasing the level of stress and anxiety consequently making them more prone to exhaustion and burnout (Australian Psychological Society 2020).

7.9 Conclusions

Although the literature had acknowledged the significance of re-evaluation of the work-life and related mental health priorities that affect the human resource productivity, policies, and procedures during the pandemic. It revealed a dramatic shift in the needs of the employees who were more concerned about work-life balance and mental health issues thus choosing to WFH. Companies were keen on re-designing physical spaces, rethinking infrastructure, and re-inventing work arrangements to reduce the cost of living and access to health care along with tackling financial results. The provision of flexibility, sustainability, and fostering work-life balance, especially for most employees (majority with parental responsibilities), affected happiness, productivity, and retention in the workplace. Additionally, this multifaceted data collation identified the emergence of a new global culture of safeguarding and promoting public health norms to protect from COVID-19. Besides physical distancing becoming an integral part of life, facemasks, hand gloves, and sanitizers were a new addition to the office and personal supplies. Unquestionably, the unprecedented situation has propelled the expansion of the simulated domain, besides the government in Australia (especially in major cities including Sydney, Melbourne, and Brisbane) took several steps to facilitate and increase contactless experience to reduce in-person contact. Meanwhile, the present situation has posed both opportunities and threats at a time. The condition of telecommuting had enabled organizations to recruit people on board without renting, leasing, or hiring office spaces. Also, employees were motivated and empowered to work remotely, with more convenient and flexible work timings (especially for those with parenting responsibilities). Contrastingly, long-term working from home might have led to increased psychological distress, diluted work-life boundaries, lack of interpersonal relationships due to increased communication gaps and lack of in-person support, diminishing organizational culture, and increased communication gap among the team members, and concerns over job security. To overcome negative consequences, the Australian government had provided grants to organizations, businesses, and service providers to maintain the level of demand and supply of services and to prioritize the actions that need to be undertaken to mitigate the impact of the downfall in this era of crisis.

This study focused on the role of technology to facilitate the ergonomics of workplaces, finding the existing gaps in gender equality, and promoting suitable and sustainable workplaces as we return to the pre-pandemic state.

Therefore, to assess, evaluate, and monitor the negative consequences of gender gaps and unequal distribution of workloads, the governments can intervene with innovative policies and procedures to reduce the economic impact for a prolonged period. Therefore, the impact of the strategies that have been adapted as an alternative to maintaining the work-life balance for both men and women may shift the dynamics of how workplaces perform and how productivity can be examined. They might work as a feasible solution to mitigate the current imbalance as aggravated during the pandemic. However, to stimulate a rewarding progress post-pandemic, organizations must acclimatize and adopt the processes of migrating to digital technology and invest in employee satisfaction through enhanced flexibility in work conditions. Although the required acceptance of the new normal might need a lot of sensitization, preparation, and simulation of the workforce to accept and invest efforts in the collaborative safeguard of organizational culture with new ways of working. This may pave a new way for leadership and team effort to manage the competency and capability during the sustained period of the pandemic. Leadership strategies during and post-pandemic could be established based on the best information, trends, and data to establish innovative approaches to problem-solving. It will stimulate proactive groundwork and planning for crisis preparedness and a significant reduction in employee burnout while enhancing the service delivery and operations output.

7.10 Limitations of the Study and Scope of Further Research

Notwithstanding the above-mentioned contributions, future research and investigation could examine and address the limitations of this study. The first limitation of this study is an inability to collect data first-hand from the employees in the workforce to analyze both quantitative and qualitative aspects of this study for prospective longitudinal studies. However, future studies can use longitudinal sampling designs and collect data from the employers, supervisors, subordinates, and workers at all various hierarchal levels of the organizations to limit method and data variance. The second limitation of this research was related to the generalizability of the results among companies/organizations (only in three major cities rather than covering the vast landscape of the Australian continent during the pandemic) therefore broadening the scope of the study rather than narrowing down the outcomes related to specific industrial/business sectors across Australian states. For generalizability of the outcomes, future investigations can collect data from various sectors for instance, the health-care sector where the workforce is required on the frontline and might not have a choice for telecommuting as others. As the immediate influence of the pandemic is pronounced, future research is implored to analyze and explore the long-term impact of the pandemic crisis on the feasibility of activities (WFH, use of public transport, etc.) and human behavior. Consequently, the research undertaken may lead to speculations and inferences about the 'new normal' and what future will unfold in upcoming years. Additionally, the use of digital strategy, transformational leadership, and remote sensing and the effect on happiness and productivity of employees due to new trial forms in the workplace could be explored further. Considerably, the future is uncertain, and the information explored and extracted could have been identified before the COVID-19 situation.

This research also sheds light on the motivations, perceptions, and resilience of employees during the pandemic. As anticipated, all research has innate limitations in its research design, methodology, data collection, and analysis. This study relied on various constructs that were adapted from other literature contributions in developed countries due to limited research in Australia as the second wave of Covid is still prevalent while this study has been undertaken. In the future, the research could be undertaken to utilize measures that were mentioned in this study and replicate the study in different contexts and regions to explore what motivates employee's performance and productivity in a post-Covid-era. Perhaps, perspective/longitudinal thematic analysis may provide more clarification on the in-depth perspectives of the employees' ideas, opinions, and beliefs on the organizational behavior, culture, and working conditions. Interpretative studies can shed more light on important insights about the employee perceptions, motivational factors, competence, autonomy, and sense of relatedness in the working spaces. There have been limited attempts to clarify the social antecedents of motivation and being resilient during dynamic and challenging situations. Also, studies on leadership, team management, co-worker support, career adaptability, and the impact of family and friends have been less examined which can further conceptualize a coping pattern during an adverse, traumatic, disaster-like event. Furthermore, motivation can make feel employees empowered to perform tasks which leads to job satisfaction and enhanced commitment to a job role. Hence, studies can examine psychological empowerment, employee resilience (especially in working parents), morale, and job engagement. Likewise, additional research opportunities can add mediator variables such as creativity, innovation, leadership, teamwork (virtual and remote teams) for examining the effect of self-efficacy on the capability and capacity of employees, and moderator variables such as job levels that may highlight the impact of extrinsic as well as intrinsic motivational factors.

7.11 Lessons Learned

Australian public sector companies have found increased connection and collaboration across teams and agencies during the pandemic as employees felt more connected across the teams (in other jurisdictions) through virtual and digital channels. Many companies found they can provide free access to almost 45% of their IT costs throughout the first wave of pandemic which further extended as the second wave of pandemic hit Australia. Organizations dedicated savings to modernizing selective technology stacks and tools (hardware and software) with no location barriers as people on the other side of the globe felt as close as those down the street. The flexibility provided through WFH options proved to narrow the focus on meeting the needs of those in crisis thus displaying the spirit of service by discovering the skills of people and capitalizing on work that gets done. Although change management has proved challenging for many organizations across major cities, it also resulted in increased resilience and motivation of the staff to work with the willingness to serve across different sectors and business units.

Additionally, the COVID-19 pandemic could be considered a wake-up call for organizations to be innovative and techno-savvy in promoting the information, communication, and reliable digital tools to accomplish organizational goals. The experimental competency and capability of the leaders has a vital role in influencing the innovative capacity and creativity output of the organizations. Therefore, future studies can investigate the effect of innovative leadership capabilities to excel during a pandemic crisis. Additionally, the initial literature review has shown the negative effects of the pandemic on 'gender inequality' with women struggling more as compared to their men counterparts. Gender differences in participation reduced labor, and decreased consumption to increase savings by women may account for a few factors that might have led to an increase in the gender gap. Although, the study has found heterogeneous effects of the pandemic on the workforce in major cities across Australia. Hence, economic and industrial policies could be tailored to support equal work-life balance among gendered roles and accelerate the economic benefits by taking the burden off one gender counterpart of the society which if neglected might hamstring future economic growth.

References

Accenture. 2021. The future of work: A hybrid work model. https://www.accenture.com/us-en/insights/consulting/future-work.

Alon, Titan, Matthias Doepke, Jane Olmstead-Rumsey, and Michèle Tertilt. 2020. "The impact of COVID-19 on gender equality." National Bureau of Economic Research.

Amis, J., and Greenwood, R. (2020). "Organisational change in a (post-) pandemic world: Rediscovering interests and values." *Journal of Management Studies*. https://doi.org/10.1111/joms.12663.

Australian Institute of family studies, Australian Government. 2021. "Two thirds of Australians are working from home." Australian Government accessed 07.01.21. https://aifs.gov.au/media-releases/two-thirds-australians-are-working-home.

Australian Psychological Society. 2020. "Maintaining employee engagement during COVID-19." Victoria, Australia: Australian Psychological Society.

Baig, A., B. Hall, P. Jenkins, E. Lamarre, and B. McCarthy. 2020. "The COVID-19 recovery will be digital: A plan for the first 90 days." McKinsey Digital accessed 10.1.22. https://www.mckinsey.com/business-functions/mckinsey-digital/our-insights/the-covid-19-recovery-will-be-digital-a-plan-for-the-first-90-days.

Barrick, Murray R. 2005. "Yes, personality matters: Moving on to more important matters." *Human Performance*. Vol. 18 (4):359–372.

Baxter, Jennifer, Megan Caroll, and Mikayla Budinski. 2020. "New report reveals how Aussie families are adjusting during COVID-19." Australian Government accessed 11.1.22. https://aifs.gov.au/media-releases/new-report-reveals-how-aussie-families-are-adjusting-during-covid-19.

Berman, Marc, and Tracy Thurkow. 2020. "Covid-19 creates a moment of truth for corporate culture." Bain & Company.

Booth, Katie. 2021. "Employers key to work/life balance during COVID-19." The University of Sydney, accessed 14.01.22. https://www.sydney.edu.au/news-opinion/news/2021/02/08/employers-key-to-work-life-balance-during-covid-19.html.

Carli, Linda L. 2020. "Women, Gender equality and COVID-19." *Gender in Management: An International Journal*. Vol. 35(7/8):647–655.

Cho, Eunae. 2020. "Examining boundaries to understand the impact of COVID-19 on vocational behaviors." *Journal of Vocational Behavior*. Vol. 119:103437. doi: https://doi.org/10.1016/j.jvb.2020.103437.

Choudhury, Prithwiraj, B.Z. Larson, and C. Foroughi. 2019. "Is it time to let employees work from anywhere." *Harvard Business Review.* Vol. 14. https://hbr.org/2019/08/is-it-time-to-let-employees-work-from-anywhere.

Cook, D., A. Rudnicka, and J. Newbold. 2021. "Work-life balance in a pandemic: a public health issue we cannot ignore." The Conversation accessed 11.01.22. https://theconversation.com/work-life-balance-in-a-pandemic-a-public-health-issue-we-cannot-ignore-155492.

Cotofan, Maria, Jan-Emmanuel De Neve, Marta Golin, Micah Kaats, and George Ward. 2021. "Work and well-being during COVID-19: impact, inequalities, resilience, and the future of work." World Happiness Report 153–190.

Demirović Bajrami, Dunja, Aleksandra Terzić, Marko D. Petrović, Milan Radovanović, Tatiana N. Tretiakova, and Abosa Hadoud. 2021. "Will we have the same employees in hospitality after all? The impact of COVID-19 on employees' work attitudes and turnover intentions." *International Journal of Hospitality Management.* Vol. 94:102754. doi: https://doi.org/10.1016/j.ijhm.2020.102754.

Forster, Bruce B., Michael N. Patlas, and Frank J. Lexa. 2020. Crisis leadership during and following COVID-19. Sage Publications Sage CA.

Hadley, Constance Noonan, Todd L. Pittinsky, S. Amy Sommer, and Weichun Zhu. 2011. "Measuring the efficacy of leaders to assess information and make decisions in a crisis: The C-LEAD scale." *The Leadership Quarterly.* Vol. 22 (4):633–648. doi: https://doi.org/10.1016/j.leaqua.2011.05.005.

Harguth, Benjamin. 2021. "Lessons from Harnessing Digital Solutions during COVID-19." Stantec Last Modified 23.06.21, accessed 10.1.22. https://www.cardno.com/news-insights/digital-bytes-lessons-from-harnessing-digital-solutions-during-covid-19/.

Hjálmsdóttir, Andrea, and Valgerður S. Bjarnadóttir. 2021. "I have turned into a foreman here at home: Families and work–life balance in times of COVID-19 in a gender equality paradise." *Gender, Work & Organization.* Vol. 28 (1):268–283. doi: https://doi.org/10.1111/gwao.12552.

Kaul, Vivek, Vijay H. Shah, and Hashem El-Serag. 2020. "Leadership during crisis: lessons and applications from the COVID-19 pandemic." *Gastroenterology.* Vol. 159 (3):809.

Khajuria, Ankur, Wojtek Tomaszewski, Zhongchun Liu, Jian-hua Chen, Roshana Mehdian, Simon Fleming, Stella Vig, and Mike J. Crawford. 2021. "Workplace factors associated with mental health of healthcare workers during the COVID-19 pandemic: an international cross-sectional study." *BMC Health Services Research.* Vol. 21 (1):262. doi: 10.1186/s12913-021-06279-6.

Kim, Rae Yule. 2020. "The impact of COVID-19 on consumers: Preparing for digital sales." *IEEE Engineering Management Review.* Vol. 48 (3):212–218.

Kinnersly, Patty. 2021. "Mark Equal Pay Day with Action to Prevent Women Being Left Behind in COVID Recovery." Our Watch, 31.08.21, Home https://www.ourwatch.org.au/resource/mark-equal-pay-day-with-action-to-prevent-women-being-left-behind-in-covid-recovery/.

Lagowska, Urszula, Filipe Sobral, and Liliane Magalhães Girardin Pimentel Furtado. 2020. "Leadership under crises: A research agenda for the post-Covid-19 Era. SciELO Brasil."

Lycett, Kate. 2021. "Australians 'remarkably resilient' during COVID-19." Deakin University accessed 18.02.22. https://www.deakin.edu.au/research/research-news-and-publications/articles/australians-remarkably-resilient-during-covid-19.

Madgavkar, Anu, Olivia White, Mekala Krishnan, Deepa Mahajan, and Xavier, Z. 2020. COVID-19 and gender equality: Countering the regressive effects. https://www.mckinsey.com/featured-insights/future-of-work/covid-19-and-gender-equality-countering-the-regressive-effects.

Mikocka-Walus, Antonina, Mark Stokes, Subhadra Evans, Lisa Olive, and Elizabeth Westrupp. 2021. "Finding the power within and without: How can we strengthen resilience against symptoms of stress, anxiety, and depression in Australian parents during the COVID-19 pandemic?" *Journal of Psychosomatic Research*. Vol. 145:110433. doi: https://doi.org/10.1016/j.jpsychores.2021.110433.

Munawar, Hafiz Suliman, Sara Imran Khan, Zakria Qadir, Abbas Z. Kouzani, and M.A. Parvez Mahmud. 2021. "Insight into the impact of COVID-19 on Australian transportation sector: An economic and community-based perspective." *Sustainability*. Vol. 13 (3):1276.

National Australia Bank. 2021. "Work from anywhere: What it means for professional services. " National Australia Bank accessed 10.1.22. https://business.nab.com.au/work-from-anywhere-what-it-means-for-professional-services-49993/.

Nguyen, Kevin. 2021. "NSW COVID case numbers hit a new high as hospitalisations eclipse last year's peak." Australian Broadcasting Corporation 04.1.22, Health. https://www.abc.net.au/news/2022-01-04/nsw-records-highest-hospitalisation-and-covid-cases/100736056.

Nilasari, B. Medina, M. Nisfiannoor, and Florensia Rosary Meida Devinta. 2021. "Changes in motivation that affect employee performance during the Covid 19 pandemic." *Jurnal Aplikasi Manajemen*. Vol. 19 (2). doi: http://dx.doi.org/10.21776/ub.jam.2021.019.02.19.

Ojo, Adedapo Oluwaseyi, Olawole Fawehinmi, and Mohd Yusoff Yusliza. 2021. "Examining the predictors of resilience and work engagement during the COVID-19 pandemic." *Sustainability*. Vol. 13 (5):2902.

Pacheco, Tyler, Simon Coulombe, Christine Khalil, Sophie Meunier, Marina Doucerain, Emilie Auger, and Emily Cox. 2020. "Job security and the promotion of workers' wellbeing in the midst of the COVID-19 pandemic: A study with Canadian workers one to two weeks after the initiation of social distancing measures." *International Journal of Wellbeing*. Vol.10 (3).

Parungao, Angelique. 2020. "The future of remote work after COVID-19: 3 common predictions." Ekoapp. com.

Queensland Department. 2022. "Queensland COVID-19 statistics." Queensland Government, accessed 05.01.22. https://www.qld.gov.au/health/conditions/health-alerts/coronavirus-covid-19/current-status/statistics.

Raghuram, Sumita. 2021. "Remote work implications for organisational culture." In Work from fome: Multi-level perspectives on the new normal, edited by Payal Kumar, Anirudh Agrawal and Pawan Budhwar, 147–163. Emerald Publishing Limited.

Ryan, Richard M., and Edward L. Deci. 2000. "Intrinsic and extrinsic motivations: Classic definitions and new directions." *Contemporary Educational Psychology*. Vol. 25 (1):54–67.

Sandoval-Reyes, Juan, Sandra Idrovo-Carlier, and Edison Jair Duque-Oliva. 2021. "Remote work, work stress, and work–life during pandemic times: A Latin America situation." *International Journal of Environmental Research and Public Health. Vol.* 18 (13):7069.

Singh, Mithilesh Kumar, Vijay Kumar, and T. Ahmad. 2020. "Impact of Covid-19 pandemic on working culture: an exploratory research among information technology (IT) professionals in Bengaluru, Karnataka (India)." *Journal of Xi'an University of Architecture & Technology*. Vol. 12 (5):3176–3184.

Spicer, André. 2020. "Organizational culture and COVID-19." *Journal of Management Studies*. Vol. 57 (8):1737–1740.

UNESCO. 2020. https://en.unesco.org/news/13-billion-learners-are-still-affected-school-university-closures-educational-institutions.

Victorian Government. 2022. "Victorian COVID-19 data." Victorian Government accessed 05.1.22. https://www.coronavirus.vic.gov.au/victorian-coronavirus-covid-19-data.

Wenham, Clare, Julia Smith, and Rosemary Morgan. 2020. "COVID-19: the gendered impacts of the outbreak." *The Lancet*. Vol. 395 (10227):846–848.

Yang, Eunhwa, Yujin Kim, and Sungil Hong. 2021. "Does working from homework? Experience of working from home and the value of hybrid workplace post-COVID-19." *Journal of Corporate Real Estate* (ahead-of-print). doi: https://doi.org/10.1108/JCRE-04-2021-0015.

Chapter 8

Rebuilding Organizations Post-Pandemic

Dipesh Ranjan
Mavenir, Singapore

Contents

DOI: 10.4324/9781003267751-11

8.1 Introduction

The business world has witnessed majorly three generations of workplace revolution, namely, the industrial revolution, the information revolution, and the current world of the social and digital revolution. Industrial revolution saw the workforce thriving for employment for survival in order to meet the fundamental needs of life. In this era, there were only a handful of ways to find out job options, and needless to mention, this workforce readily accepted and succumbed to higher management demands.

However, with the information revolution, the age of 'Boss is Always Right' came to an end. The grand 'Information Revolution' era, where employees did not come to work to fulfil their basic needs of food and shelter (as their parents/grandparents did), rather strived for a high standard of living. For example, the workforce joined organizations offering handsome salaries, especially in the private sector that helped them to meet their esteem needs like home, cars, children's education; this was the dream of the workforce from the industrial revolution era which could be met only after retirement in their late 50s or 60s.

On the other hand, the 'Social/Digital Revolution' brought a sea change in the perspective and lifestyle of gen x, y, and z. The current workforce has shifted the focus from esteem needs to building good 'quality of life.' The race is for quality in terms of job/role/opportunity, organizational environment/culture, learning and reward, and so on and so forth. Googling is the buzzword today, learning has taken a massive shift through YouTube, and job hunting has taken a U-turn with LinkedIn. Consequently, businesses have realized the dire necessity to reinvent themselves with the growth of the social/digital revolution. Rifkin (2011), in his book on the third industrial revolution, has emphasized the term 'sharing economy' and believes in the use of conservation of energy by the digitization process. This shared economy will definitely play a vital role in shaping the business world post-pandemic.

The need to rebuilding organizations has taken a leap of faith ever since the human race faced the biggest pandemic ever in the century, COVID in 2019. The surge in COVID in the past few years has led to a 'big resignation tsunami' and the focus of the workforce has shifted toward the quality of healthy life. This shift is compelling organizations to manage their talented workforce and build a healthy workplace along with focusing attention on operating in the marketplace for bottom lines. The current shift in this era, therefore, required visionary leadership.

8.2 Life after Pandemic in Organizations

The good news is that COVID cases are subsiding and most of the population across the world has had their doses of vaccinations. The employees are returning to their older office regime. However, there is certain apprehension among the employees to return to work with 100 percent occupancy in the office building due to the safety concerns. Organizations are coping with the issue, and many will do so in the future for one simple reason: the rate of attrition is an all-time high. Rather than addressing the root causes of attrition, many organizations resort to well-intended short-term fixes that fail. These fixes may include an increase in compensation or financial benefits or awarding 'thank you' bonuses without making serious efforts to build employees' relationships with coworkers and bosses. Therefore, it is often observed that employees perceive such rewards as transactions instead of appreciation.

Post-COVID, organizations have realized that their workforce is burned-out and exhausted, and many are in mourning after losing their dear ones. The workforce has a desire to rediscover and revise the sense of purpose in their employment. They also seek to form social and interpersonal bonds with their coworkers, subordinates, and bosses. It goes without saying that they want compensation, benefits, and perks; however, it is also pertinent that they want to feel appreciated by management. Some employees often take intentional decisions to leave regular full-time employment since many organizations neglect to invest in a satisfying employee experience at the workplace. There can be several cogent reasons for an increase in employee voluntary turnover post-pandemic. A few pertinent ones are as follows:

i. **Pandemic burnout:** stress and uncertainty about the job and a sense of isolation leading to feeling disconnected from organizational culture and values.
ii. **Re-evaluated priorities:** several months of not being able to lead a 'normal' life has created a determination for many to change things for the better rather than just 'wait and see,' often moving to an entirely new career.

iii. **Work-from-home:** working from home has provided greater flexibility and freedom. Many will never go back to working in an office full time.
iv. **De-motivated and undervalued by managers**: feeling valued, supported, and worthy is more important than money.
v. **Lack of growth:** due to the lack of engagement in worthy projects from managers and a lack of learning and development opportunities for career enhancement, many (particularly Gen Z) are driving their own growth via the increased availability of online learning resources.
vi. **Online meetings:** increased online meetings on Zoom/MS Teams/ Google meet, etc., mixing of personal and professional life has resulted in longer working hours.
vii. **Financial independence:** increased savings due to reduced expenditure on the daily commute and social events have provided financial freedom to move jobs.

8.3 Facets of Rebuilding in Organizations

8.3.1 Rebuilding Workforce

The widespread impact of COVID will permanently alter the nature of labour in a wide range of professions. The borders of the 'new normal' are increasingly being questioned by businesses. COVID has brought about such broad and widespread changes that they will continue to influence how people work long after the epidemic has passed. As Winston Churchill phrased it, we'll be at the 'end of the beginning.' Organizations are expected to consider workplace safety measures and their preparedness for unique circumstances like COVID. The employees' health and well-being are regarded as a top priority. The new normal will emerge as a result of changes in the outlook of organizations toward their workforce and their relationship with management. Let's explore a way forward to the new normal.

8.3.2 Employee Engagement

Consistent and acceptable pay, as well as job stability, are important factors for retaining talent in the organizations but employee engagement is a cognitive and rational factor (Kular et al. 2008) that goes a long way in achieving organizational objectives. Employees who connect with their professions and operate in a healthy work environment are able to satisfy their sense of belongingness while also laying the groundwork for engagement. Recognition, appreciation for efforts, and a sense of importance in the workplace help to build self-esteem and increase employee engagement in challenging projects. Another important aspect is the alignment of employees' personal objectives with corporate values and opportunities for growth which provide the foundation for self-actualization in a high-engagement workplace, leading to increased productivity and performance.

8.3.3 Employee Well-Being

More often than not, an important question hovers around the workforce mind, 'Am I being taken care of?' Organizations should work on answering this question for talent management purposes. Post-COVID, heavy salary cuts and pay-check delays have resulted in undermining the standards of employee well-being. Organizations should place a premium on employee health and make sure that basic needs like food, housing, and other living expenses are addressed, regardless of income fluctuations.

8.3.4 Job Security

Job security has been identified as one of the most significant stresses in the face of a pandemic, offering a direct threat to the second rung of Maslow's hierarchy of needs (McLeod 2007). Employees are terrified of losing their employment, and this fear makes them work longer hours sacrificing their personal life. Organizations may bridge this gulf by devising ways. Rather than abruptly ending, it is critical to maintain open channels of contact with the workforce and to educate them on the present financial position of the company on a regular basis. Indeed, effective communication is preferable to ambiguity and confusion among employees regarding their strategic position in the organization.

8.3.5 Need for Belongingness

Brooks and Ochlan (2015) of Harvard University has very well described 'Weighing the risk of loneliness.' Often employees ask the question, 'do I belong here?' Many organizations are still operating 'work-from-home' arrangements for their employees even when pandemic has become endemic. Needless to mention, loneliness leads to a high rate of employee burnout, turnover, and disengagement.

While some businesses are fighting to make ends meet as a result of the current crisis, others are battling with the third hierarchy of needs. When employees are in a physical area and among their peers and coworkers, they feel a sense of belonging. Employees are dissatisfied with their employment and crave a sense of belonging in the face of long-distance work. To fight this, businesses must improve their communication skills and encourage virtual engagement, as well as team bonding events. Simultaneously, outside work, virtual coffee, and chat meetings may assist in reigniting that sense of belongingness and involvement.

8.3.6 Empathy

When the pandemic subsides, the lives of the workers are not completely restored back to normal. Some of the employees are going back to work without a loved one who passed away as a result of the COVID. Throughout the epidemic, it is important to lead the workforce with compassion and kindness. Being sympathetic does not always imply having a soft corner rather the managers need to be creative and innovative when it comes to spreading compassion. Leaders must also be fair to them and seek help if their mental health is to be preserved.

8.3.7 Self-Esteem

Another question pondering the minds of employees is 'Am I respected in my organization?' Companies that have mastered communications have advanced to the next level in Maslow's hierarchy of demands. The more a firm celebrates and honours its employees in a conventional setting, the more motivated and engaged they are at work. Remote work, as well as increased unpredictability and the need to adapt to new working styles, has pushed performance recognition to the back burner. Employees who perform well as well as those who perform poorly are treated equally, thereby, resulting in a lack of motivation among high-performers and achievers. Organizations may address this issue by providing digital rewards and appreciation to their employees by appraising their performance. The need for esteem in employees is majorly fulfilled in the most forward-thinking businesses, which go to great lengths to recognize and reward their employees' achievements, even if it's virtual.

8.3.8 Nirvana in Corporate Life

Getting Nirvana (Self-actualization) is at the summit of the hierarchy of needs (McLeod 2007). Due to conflicting agendas and a lack of a healthy work environment, people, on the other hand, are unable to realize their full potential. Businesses must guarantee that the employment they supply to their workers contributes to their personal and professional growth in order to close this gap. Actualization can be achieved by providing a virtual learning platform to the employees and this can be realized by creating a customized avatar. Even online mentoring and coaching can go a long way in achieving Nirvana.

8.3.9 Diversity, Equity, and Inclusion (DEI)

It goes without saying that DEI (Loden 1996) is a challenge faced by most organizations and a considerable amount of efforts are being taken by organizations to sensitize their managers regarding the same. Employers may unknowingly create imbalances on various dimensions, including salary, benefits, and career, as they plan for new ways of working. Varied employee categories – women workforce, caregivers, coloured workforce, parents who work full time, an employee with low income, etc. – have diverse effects and different experiences as a result of the epidemic. Groups that may be subjected to inequitable treatment due to workplace unconscious prejudice should not be neglected. Employers should assess their compensation packages and gain a better grasp of worker preferences and the effects of present benefits and provisions. Organizations also need to train and sensitize managers to be aware of these concerns and to make sure they're leading in the right direction. Chief resilience officers and chief diversity and inclusion officers are two examples of emerging organizational roles.

8.3.10 Rebuilding Workplace

The traditional workplace was destroyed by COVID; therefore, a pertinent question arises, what should employers do now? As we enter a new post-pandemic normal, how businesses balance their sustainability with their employees needs to be analyzed. Organizations should explore new options as they rethink and reimagine how work gets done during the endemic to help them cope with the challenges of hybrid organizations and the returning to the office premises. The organizations should embrace the new world, accept the new demands, and adopt a model which focuses on basic fundamental human needs.

8.3.11 Innovate the Physical Working Environment

Working from any location, at any time, is becoming a viable option for non-location-based tasks. Work pods in neighbourhoods, for example, might be merged with typical offices. The post-pandemic conclusion is self-evident: a hybrid work paradigm in which a segment of the workforce spends time outside the traditional workplace. 'Who has to be at work, when, and why?' is the most critical question. Employees want to know what their work arrangements will be in the future. Organizational leaders are responsible for charting the way for managers and employees. Transparent and frequent communication, with management playing a key role, may aid in ensuring that the company runs well. In order to create informal social contacts in distant workplaces, it's also vital to organize virtual watercoolers.

8.3.12 Organizational Policies

In recent online surveys, employee well-being, and productivity were shown to be greater in firms that detailed more specific policies and plans for the future workplace. Employees in firms that have appropriately created post-COVID-19 work arrangements reported a twofold rise in sentiments of support, a threefold increase in feelings of inclusion, and a nearly fivefold increase in reported feelings of individual productivity. Trying to apply a one-size-fits-all method to the workforce can backfire, particularly for women, individuals with lower socioeconomic standing, and people living in developing nations.

8.3.13 Workplace Changes

Hybrid work will continue to exist: employees want to return to work, but not on a daily basis. They also want the option of working how and where they choose. It's important to look at how we now define an 'office.' Working in an office in the pre-pandemic era required driving or using public transit to a location known as the office. Employees sent an email with the subject 'away from the office' if they left the office. During the past two years, the attitudes around employment have evolved. 'Office' is a word now commonly used to denote a work environment. The office will continue to exist, but with an objective where the workplace will have a new shape promoting healthy collaboration of teams, enhancing learning and creativity among the employees working in solidarity.

These changes in the workplace call for digital intervention using technical collaboration tools; reaping the benefits of robotics and artificial intelligence in order to augment the growth of virtual reality. Areas of advancement in areas pertaining to analytics, blockchain, cryptocurrency, and the like will give a new look to the organizations.

8.3.14 Employee Experience

Employers should rethink how and where work is done by utilizing a proactive hybrid work strategy thereby enabling high-quality employee experience. If done effectively, this can lead to better employee engagement prospects and boost productivity at all levels of the organization. The C-suite of the organizations should focus on making the employees' work exciting, captivating, and engaging, whether it is done at the office or not (Morgan 2017).

8.3.15 Building Supervisory Role

Needless to mention, the supervisors try to achieve their organizational goals by aligning the key performance indicators with the strategic mission and vision of the company, ensuring that jobs are well designed. However, this role of supervisor has undergone change and needs lifting too. In the case of a business environment that is hybrid, supervisors will not devote time to monitor operations that much as in the case of a physical environment; more time now can be dedicated to communicating and mentoring the employees.

8.3.16 Corporate Culture

Culture eats strategy in breakfast: therefore, organizations should build culture as their first strategy. Culture is the sum of people's thoughts and beliefs, as well as their behaviours, facilitators, and experiences, in an organization. Businesses need to be sensitive in defining and building their corporate culture in order to meet the challenges and demands of the disruptive future. The issue is recognizing – and retaining – the beliefs and philosophy of existing corporate culture that help build a successful organization, while letting go of certain sets of cultural aspects that lead to a setback (Schein 2009).

8.3.17 Performance Management Model

As more individuals work from home, managers may be unsure of how to assess and measure employee performance and productivity. By viewing and watching someone before the outbreak, we knew they were working properly, and their evaluation was based on their observed behaviour. Post-pandemic the argument for performance has evolved, with an emphasis on different attributes. The emphasis will be on effect and value generation rather than milestones and targets.

8.4 Conclusion

There is an emerging interest in organizations to rebuild their workforce and workplace in line with the needs of the dynamic business environment. Needless to mention, there is a considerable shift in practices, policies, and experiences caused by the pandemic that may lead to an organizational design best suited for the optimal utilization of resources (Amis and Greenwood 2020).

Leaders should concentrate not just on assisting the company in achieving its goals, but also on ensuring that the group can use its collective strengths and practices in order to achieve organizational objectives. It is evident that working from home is best suited for jobs that involve some autonomy and can be readily defined and delivered over great distances. When activities are interconnected, sharing knowledge and information in an effective manner is crucial, which can be done through physical mode. Emphasis should be laid on job requirements and not the organizational preferences and individual discretion should be allowed when people believe that a hybrid strategy will not work.

In the present times of a dynamic work environment, no option is fool-proof and complete to achieve success for the organizations. In the new normal, rethinking the techniques in accordance with the hierarchy of needs and demands may boost employee engagement to new heights. Attention must be paid to the human side of the business, as well as a reaffirmation of goals and values, in order to emerge effectively post-pandemic with top talent retention. Perhaps, when it's too late, corporations should use recognition-rich recommitment ceremonies (akin to wedding vow renewal) instead of departure parties. Creativity is vital for improving interpersonal interactions and fostering a sense of belonging in present times.

References

Amis, J., & Greenwood, R. 2020. Organizational change in a (post-) pandemic world: Rediscovering interests and values. *Journal of Management Studies*, Vol. 58 (2): 582–586. Doi:10.1111/joms.12663.

Brooks, A. C., & Ochlan, P. J. 2015. *The conservative heart*. Blackstone Publishing.

Kular, S., Gatenby, M., Rees, C., Soane, E., & Truss, K. 2008. Employee engagement: A literature review. Kingston University, Kingston. https://eprints.kingston.ac.uk/id/eprint/4192/1/19wempen.pdf; accessed on 12th January 2022.

Loden, M. 1996. *Implementing diversity*. Irwin Professional.

McLeod, S. 2007. Maslow's hierarchy of needs. *Simply Psychology*, Vol. 1: 1–18.

Morgan, J. 2017. *The employee experience advantage: How to win the war for talent by giving employees the workspaces they want, the tools they need, and a culture they can celebrate*. John Wiley & Sons.

Rifkin, J. 2011. *The third industrial revolution: how lateral power is transforming energy, the economy, and the world*. Macmillan.

Schein, E. H. 2009. *The corporate culture survival guide*. Vol. 158. John Wiley & Sons.

Index

Note: Locators in *italics* represent figures and **bold** indicate tables in the text.

W

Printed in the United States
by Baker & Taylor Publisher Services